FLEXIBLE TEST AUTOMATION

FLEXIBLE TEST AUTOMATION

A SOFTWARE FRAMEWORK FOR EASILY DEVELOPING MEASUREMENT APPLICATIONS

PASQUALE ARPAIA,
ERNESTO DE MATTEIS, AND
VITALIANO INGLESE

MP MOMENTUM PRESS

MOMENTUM PRESS, LLC, NEW YORK

Flexible Test Automation: A Software Framework for Easily Developing Measurement Applications
Copyright © Momentum Press®, LLC, 2015.

First published by Momentum Press®, LLC
222 East 46th Street, New York, NY 10017
www.momentumpress.net

ISBN-13: 978-1-60650-383-6 (print)
ISBN-13: 978-1-60650-385-0 (e-book)

Momentum Press Automation and Control Collection

DOI: 10.5643/9781606503850

Cover and interior design by Exeter Premedia Services Private Ltd., Chennai, India

10 9 8 7 6 5 4 3 2 1

Printed in the United States of America

ABSTRACT

In laboratory management of an industrial test division, a test laboratory, or a research center, one of the main activities is producing suitable software for automatic benches by satisfying a given set of requirements. This activity is particularly costly and burdensome when test requirements are variable over time. If the batches of objects under test have small size and frequent occurrence, the activity of measurement automation becomes predominating with respect to the execution.

In this book, the development of a software framework is shown to be as a useful solution to satisfy this exigency. The framework supports the user in producing measurement applications for a wide range of requirements with low effort and development time. Furthermore, the software quality, in terms of flexibility, usability, and maintainability, is maximized.

After a background on software for measurement automation and the related programming techniques, the structure and the main components of a software framework for measurement applications are illustrated. Their design and implementation are highlighted by referring to a practical application: the Flexible Framework for Magnetic Measurements (FFMM) at the European Organization for Nuclear Research (CERN). Finally, an experimental approach to the software flexibility assessment of measurement frameworks is presented by highlighting its application to FFMM.

KEYWORDS

application software, automatic programming, magnetic measurements, measurement automation, particle accelerators, software frameworks, software systems

CONTENTS

LIST OF FIGURES   xi

LIST OF TABLES   xvii

SUMMARY   xix

ACKNOWLEDGMENTS   xxi

CONVENTION ABOUT THE NOTATION   xxiii

INTRODUCTION   xxv

PART I   BACKGROUND   1

1 SOFTWARE FOR MEASUREMENT APPLICATIONS   3

  1.1 Overview   3

  1.2 Basics   3

  1.3 Main Market Solutions   5

  1.4 Research: State of the Art   14

  References   23

2 SOFTWARE FRAMEWORKS FOR MEASUREMENT APPLICATIONS   33

  2.1 Overview   33

  2.2 General Concepts   33

  2.3 Why a Framework for Measurements?   36

  2.4 Domain Specific Languages   38

  2.5 Requirements of a Framework for Measurement Applications   41

  References   43

3 OBJECT- AND ASPECT-ORIENTED PROGRAMMING FOR MEASUREMENT APPLICATIONS   **47**

    3.1 Overview   47

    3.2 Object-Oriented Programming   47

    3.3 Aspect-Oriented Programming   56

    References   62

PART II   METHODOLOGY   **65**

4 A FLEXIBLE SOFTWARE FRAMEWORK FOR MEASUREMENT APPLICATIONS   **67**

    4.1 Overview   67

    4.2 Framework Paradigm   68

    4.3 *Fault Detector*   79

    4.4 *Synchronizer*   85

    4.5 Measurement-Domain Specific Language   95

    4.6 Advanced Generator of User Interfaces   101

    References   109

5 QUALITY ASSESSMENT OF MEASUREMENT SOFTWARE   **115**

    5.1 Overview   115

    5.2 Software Quality   115

    5.3 The Standard ISO 9126   120

    5.4 Quality Pyramid   122

    5.5 Measuring Flexibility   126

    References   128

PART III   CASE STUDY   **131**

6 THE FLEXIBLE FRAMEWORK FOR MAGNETIC MEASUREMENTS AT CERN   **133**

    6.1 Overview   133

    6.2 Methods for Magnetic Field Measurements   134

    6.3 Automatic Systems for Magnetic Measurements   139

    6.4 Software for Magnetic Measurements at CERN   140

    6.5 Flexibility Requirements for Magnetic Measurement Automation   142

6.6   The Framework FFMM   146

References   165

7   IMPLEMENTATION   **171**

7.1   Overview   171

7.2   Base Service Layer   172

7.3   Core Service Layer   187

7.4   Measurement Service Layer   191

7.5   User Service Layer   198

7.6   Software Quality Assessment   209

References   222

8   FRAMEWORK COMPONENT VALIDATION   **225**

8.1   Overview   225

8.2   *Fault Detector*   226

8.3   *Synchronizer*   236

8.4   Domain Specific Language   241

8.5   Advanced User Interfaces Generator   247

References   249

9   FRAMEWORK VALIDATION ON LHC-RELATED APPLICATIONS **251**

9.1   Overview   251

9.2   On-Field Functional Tests   252

9.3   Flexibility Experimental Tests   272

9.4   Discussion   277

References   278

INDEX   **281**

LIST OF FIGURES

Figure 1.1.   Main market solutions tree: leaders and products.   7

Figure 1.2.   Plot of technical product categories versus producers.   13

Figure 1.3.   State of art research tree: software for measurement
applications.   15

Figure 2.1.   Test engineer and application user exploit a
measurement software framework.   43

Figure 3.1.   Architecture of a simple measurement and control
system: (1) a first event triggers the model (*abstract
factory*), (2) the model produces a system software
instance (*measurement item*), (3) the instance drives the
sensor to carry out a measurement, (4) the sensor returns
back the reading, (5) the model instance processes the
data, and (6) drives the actuator suitably.   55

Figure 3.2.   Example of a crosscuttings concern in measurement
software: The fault detection.   57

Figure 3.3.   (a) Code scattering and (b) tangling.   58

Figure 3.4.   An example of a straightforward AOP program
implemented in AspectJ.   60

Figure 4.1.   Working principle of a framework for measurement
applications.   70

Figure 4.2.   Layered architecture model of the SFMA.   71

Figure 4.3.   Architecture of a SFMA.   72

Figure 4.4.   The UML model of the framework kernel.   73

Figure 4.5.   UML diagram of the multilayered architecture of the
framework (for the sake of simplicity the *User service*
layer is not reported).   74

Figure 4.6.   Measurement framework event handling architecture.   76

Figure 4.7.   Measurement framework actions and listeners infrastructure.   77

Figure 4.8.   Logger architecture of the framework.   78

Figure 4.9.   An excerpt of the hierarchy of the *Fault Detector*.   83

Figure 4.10.   Levels of faults interception.   83

Figure 4.11.   Fault notification publish-subscribe architecture.   84

Figure 4.12.   Working example of the *Execution Graph*.   89

Figure 4.13.   Code lines for the *Execution Graph* definition.   90

Figure 4.14.   Example of a Petri net.   90

Figure 4.15.   Architecture of *Synchronizer*'s classes.   92

Figure 4.16.   Implementation of the Execution Graph entities (nodes and arrows).   93

Figure 4.17.   A generic *Task Manager* uses the *Synchronizer* to select an executable task.   94

Figure 4.18.   A generic task manager uses the *Synchronizer* to trace the change of the tasks execution status.   95

Figure 4.19.   MDSL process according to the pattern "source-to-source".   98

Figure 4.20.   Semantic model and MDSL architecture.   99

Figure 4.21.   *Model-Viewer-Interactor* (MVI) approach.   105

Figure 4.22.   Example of (a) a View model and (b) its final aspect.   106

Figure 4.23.   View XML description example.   107

Figure 4.24.   MDSL script example.   108

Figure 4.25.   *Abstract factory* pattern for the GUI engine.   109

Figure 5.1.   The ISO 9126 quality model.   121

Figure 5.2.   Approaches to software quality according to ISO 9126.   122

Figure 5.3.   The three major aspects quantified by the overview pyramid.   123

Figure 5.4.   Example of a completed overview pyramid (the metrics values refer to an application example).   123

Figure 6.1.   Rotating coils measurement principle.   136

Figure 6.2.   Layout of the rotating-coil based measurement system controlled by MMP.   141

Figure 6.3.   An excerpt of the hierarchy of the *Fault Detector*.   147

Figure 6.4.   Levels of faults interception.   149

Figure 6.5.   Fault notification architecture "publish-subscribe"
for the device *EncoderBoard*.   150

Figure 6.6.   Device creation/destruction interception.   151

Figure 6.7.   Collaboration scenario started by *createDevice*
interception on *EncoderBoard*.   152

Figure 6.8.   Sequence diagram showing the detection of a wrong
parameter configuration of an *EncoderBoard* instance. 153

Figure 6.9.   AOP-based architecture of a *Synchronizer* for an
automatic measurement system.   155

Figure 6.10. MDSL implementation in FFMM.   157

Figure 6.11. Specific grammar example of MDSL in *Xtext*.   158

Figure 6.12. Domain-specific Builder rules of MDSL in *Xpand*.   159

Figure 6.13. Main window.   161

Figure 6.14. Code for the *FdiClusterFrame*.   161

Figure 6.15. *FdiClusterFrame* window.   162

Figure 6.16. Code for the *EncoderFrame*.   162

Figure 6.17. *EncoderFrame* window.   163

Figure 6.18. Code example for the *GIC*.   163

Figure 6.19. Windows generated by the *GIC*.   164

Figure 6.20. Code for the plot function.   164

Figure 6.21. Code for the FDI data plot function.   165

Figure 6.22. Plot window.   165

Figure 7.1.   Structure of the configurator *CommunicationBus*.   172

Figure 7.2.   Structure of the communication services.   174

Figure 7.3.   Active component design for logging infrastructure. 176

Figure 7.4.   Diagram of the class *Transducer*.   177

Figure 7.5.   Circuit schematic of the physical cryo-thermometer
RTD, CERNOX CX.   179

Figure 7.6.   Diagram of the class *MotorController*.   181

Figure 7.7.   Acceleration (a) and velocity (b) profiles.   182

Figure 7.8.   Wizard to start new *Xtext* project.   201

Figure 7.9.   DSL grammar.   203

Figure 7.10. Generate *Xtext* artefacts.   203

Figure 7.11. Deployment of the DSL plug-ins.   204

Figure 7.12. *Xpand* template.   205

Figure 7.13. Example of assignment operators in FFMM project. 206

Figure 7.14. Entity. 207

Figure 7.15. *Abstract type* rule. 207

Figure 7.16. *Token* rule expressed. 208

Figure 7.17. Comments. 208

Figure 7.18. DSL test engineer steps. 209

Figure 7.19. Assistance to the measurement procedure. 209

Figure 7.20. The part of the Script in C++. 210

Figure 7.21. The same Script of Figure 7.20 in DSL. 211

Figure 7.22. ISO 9126 subcharacteristics in FFMM 3.0 (0 indicates the best quality level). 214

Figure 7.23. ISO 9126 characteristics in FFMM 3.0 (0 indicates the best quality level). 215

Figure 7.24. Overview Pyramid for the FFMM 3.0 source code. 215

Figure 8.1. Layout of the rotating coil measurement setup. 228

Figure 8.2. The abstract *FaultDetector* aspect. 230

Figure 8.3. Excerpt of *DigitalIntegrator_FaultDetector*. 230

Figure 8.4. Percentage lines of code (LOC%) of fault detection concern in device modules for OOP and AOP versions. 232

Figure 8.5. (a) DOS and (b) DOF comparisons of OOP and AOP versions with respect to Fault Detection concern. 232

Figure 8.6. (a) Total average and (b) worst case overhead times spent in aspect runtime. The pointcut expressions numbering refers to Table 8.2. 236

Figure 8.7. Layout of the split-coil permeability measurement setup. 237

Figure 8.8. Current cycles. 238

Figure 8.9. FFMM script fragment defining the *Execution Graph*. 238

Figure 8.10. Execution graph of the case study on permeability measurement. 240

Figure 8.11. Hysteresis curve of the material. 241

Figure 8.12. Superconducting magnet test script. 242

Figure 8.13. Measured normal sextupolar "*decay*" and "*snapback*" as a function of (a) the time and (b) as a function of the measured current for different supply current cycles (data are scaled to be compared). 244

Figure 8.14. Permeability measurement MDSL script. 245

Figure 8.15. Permeability measurement results for different current
ramp rates: (a) 0.5 A/s and (b) 0.01 A/s. 246

Figure 8.16. FDI configuring forms. 248

Figure 8.17. A window plotting some current cycles. 248

Figure 8.18. Relative permeability versus magnetic field curve. 249

Figure 9.1. Split-coil permeameter. 253

Figure 9.2. (a) Architecture and (b) experimental setup of the
permeability measurement bench at CERN. 255

Figure 9.3. MDSL script for permeability measurement. 257

Figure 9.4. Measured current and computed magnetic field without
sample. 258

Figure 9.5. First magnetization curve of the soft steel sample. 259

Figure 9.6. Relative permeability of the soft steel sample. 259

Figure 9.7. Architecture (a) and experimental setup (b) of the
automatic measurement station based on rotating coils
at CERN. 261

Figure 9.8. Superconducting magnet test script. 263

Figure 9.9. Measured sextuple component b_3 versus (a) current
and (b) time, in units (10^{-4} fraction of the main field
component). 264

Figure 9.10. Computed MSCs powering current cycle for sextupole
compensation. 267

Figure 9.11. Architecture of the tracking test measurement station. 268

Figure 9.12. LHC standard current cycle. 269

Figure 9.13. DSL script for rotating coil-based measurement. 270

Figure 9.14. Integral b_3 component versus current with and without
compensation, in the dipole magnet *MB2524* during
an LHC cycle. 271

Figure 9.15. Residual integral b_3 component versus current with
compensation, in the dipole magnet *MB2524* during
an LHC cycle. 272

Figure 9.16. Estimation of the sextupole with the old standard
and the FFMM platform (FAME). 273

LIST OF TABLES

Table 2.1. Main software characteristics and users they address   43

Table 3.1. Classification of patterns   53

Table 5.1. Metrics catalog   117

Table 5.2. Metrics catalogue (NDD, AHH, DOF, and DOS are proportions)   119

Table 6.1. Main software characteristics and users they address   145

Table 7.1. Computation of the sensor resistance R_{thm} 180

Table 7.2. Complexity and Object-Oriented metrics with their target values   212

Table 7.3. FFMM 3.0 size metrics summary   212

Table 7.4. FFMM 3.0 complexity metrics   213

Table 7.5. FFMM 3.0 Object-Oriented metrics   213

Table 8.1. Fault detection code in each device module and computation of percentage DOF and DOS metric for both OOP and AOP versions (OOP: Object-Oriented Programming; AOP: Aspect-Oriented Programming; LOC: Lines of code; DOF: Degree of focus; DOS: Degree of scattering)   231

Table 8.2. Worst average times spent in aspect runtime with respect to device creation and destruction and fault detection point cuts   235

Table 9.1. Injection harmonic tolerance (in hundreds of ppm of the main dipole)   260

Table 9.2. Generalized evolutions cost metric for different classes of changes in FFMM   274

Summary

*What we need, then, is a new "paradigm"—a new vision of reality;
a fundamental change in our thoughts, perceptions, and values.
The beginnings of this change, of the shift from the mechanistic to the
holistic conception of reality, are already visible in all fields and are
likely to dominate the present decade.... The purpose of this book is to
provide a coherent conceptual framework that will help them recognize
the communality of their aims. Once this happens, we can expect the
various movements to flow together and form a powerful force for social
change. The gravity and global extent of our current crisis indicate
that this change is likely to result in a transformation of unprecedented
dimensions, a turning point for the planet as a whole.*
—Fritjof Capra, The Turning Point: Science,
Society, and the Rising Culture

This book covers the specification, design, prototyping, and validation of
a software framework for supporting the development of programs for
measurement control and data acquisition. It is completed also by a true
case study on an actual software package developed at the European Organization for Nuclear Research (CERN) in cooperation with the University
of Sannio: the *Flexible Framework for Magnetic Measurements* (FFMM).
FFMM is currently in use at the section of Magnetic Measurement of
CERN and constitutes the software part of the new platform for magnetic
measurements, including also high-performance hardware.

In laboratory management, one of the main activities is producing
suitable software for automatic benches in satisfying a given set of test
requirements. This activity is particularly costly and burdensome when
test requirements are variable over time. When a batch of objects to be
tested arrives, if the test is burdensome to be carried out manually, the
need for developing an automatic bench arises. Test engineers define the
measurement requirements, and the automatic bench is designed and
developed. If the batches of objects have small size and frequent occurrence, the activity of measurement automation becomes predominating
with respect to the test execution. Most significant efforts are devoted to
instrumentation interfacing and software development.

A software framework for measurement applications is conceived as a unified solution to drive all the existing and future park of measurement systems in a generic field for an industrial test division, a test laboratory, or a research center. It allows the easy development of software for measurement and test applications under highly and fast-varying requirements. The framework supports the user in producing measurement applications for a wide range of requirements with low effort and development time. As a matter of fact, the development effort is reduced and finalized, by relieving the test engineer of development details. Furthermore, a framework allows the software quality, in terms of flexibility, usability, and maintainability, to be maximized. The framework can be configured for satisfying a large set of measurement applications in a generic field for an industrial test division, a test laboratory, or a research center.

In this book, the development of a software framework for measurement and test applications is illustrated in order to reach these goals by addressing the aforementioned issues of laboratory management. The framework exploits (a) *Object-Oriented Programming* (OOP) and (b) an innovative technology, the *Aspect-Oriented Programming* (AOP). AOP extends the Object-Oriented paradigm in order to encapsulate features transversal to several functional units (*crosscutting concerns*) by means of new software modules, the *aspects*.

A software framework for measurement and test applications includes utilities for (a) fault detection, (b) software synchronization, (c) automatic generation of user interfaces, and (d) a *Measurement Domain Specific Language* to provide the test engineer with an easy and fast way to write measurement scripts containing formal descriptions of the test protocols.

The framework is designed to be flexible, maintainable, reusable, and efficient. To assess the fulfillment of these project goals, the internal quality of its source code is to be assessed by means of suitable metrics, according to the reference model defined in the standard ISO 9126.

Finally, an experimental approach to the software flexibility assessment of measurement frameworks is presented by highlighting its application in the context of FFMM. The effectiveness of this approach is proven by reporting in this book the tests carried out on the field at CERN with different protocols and measuring equipment. The framework's effectiveness is evidenced in the development of software for measurements with very different requirements. In the first five years of FFMM operation, increasing grades of flexibility were surveyed, by moving from programming to user script level, from the point of view of both the developer and the test engineer.

ACKNOWLEDGMENTS

This book originates from the PhD Thesis of Vitaliano Inglese, supervised by Prof. Pasquale Arpaia of the University of Sannio, and defended at the Department of Electrical Engineering of the University of Naples Federico II, under the Tutorship of Prof. Nello Polese and Dr. Marco Buzio, which the Authors acknowledge gratefully.

The PhD Thesis is based on research work carried out mainly at CERN in Geneva, Switzerland, under the framework of two collaborations between the University of Sannio and the CERN in 2007 and in 2010, whose support the Authors acknowledge gratefully.

The Authors would like to express their gratitude to all the persons who, in different ways, have contributed to the realization of this book. Their help and support from the first elaboration of the idea of a framework for measurement applications to the direct involvement in the design and implementation phases, and their comments and proofreading during the long process of writing this book were precious to the Authors.

First of all, the Authors would like to thank Felice Cennamo for inspiring their research work as a whole and for convincing Pasquale Arpaia to start his first leave at CERN. Alessandro Masi deserves a particular mention, both for originating the Authors' work at CERN, and for his timeless good mood.

Authors would like to thank Nello Cimitile, at that time Rector of the University of Sannio, and Pasquale Daponte, head of LESIM, for granting special permission allowing Pasquale Arpaia to work at CERN for the past nine years. Nello understood the strategic importance of such a work seven years before the CERN worldwide resonance due to the Higgs boson finding.

A special thank is given to Philippe Lebrun, head of the "joyful war machine" for research and development that was the Accelerator Technologies (AT) Department at CERN. AT was a free forge of ideas that had a determining role in the growth of the Authors, as well as, a little bit more importantly, in the success of the Large Hadron Collider.

This book would never have been conceived without Luca Bottura's special ability of looking ahead into the future, from whom came the first idea of launching the research project of the Flexible Framework for Magnetic Measurements (FFMM) at CERN.

The Authors would like to thank Louis Walckiers, head of the CERN Magnetic Measurements Group for his friendly management of the project FFMM, and Marco Buzio for his help and encouragement in the development of the idea of a framework for magnetic measurements. A fruitful role was played by Laurent Deniau that beset the Authors with all his suggestions and patience in conceptual discussions.

The Authors would really like to thank Giuseppe Di Lucca and Mario Luca Bernardi of the University of Sannio, whose collaboration was precious for the first refinement of the framework core architecture, and, above all, for the introduction of the Aspect-Oriented approach to the design of important components of the framework.

The Authors are also grateful to Giovanni Spiezia, Stefano Tiso, Domenico Della Ratta, Giancarlo Golluccio, Giuseppe Montenero, Carlo Petrone, Lucio Fiscarelli, Fabio Corrado, Giuseppe La Commara, Felice Romano, and Juan Garcia Perez who remarkably contributed to the definition of the framework core and components, and for their implementation in the FFMM at CERN.

Authors would also like to thank Walter Scandale for his constant encouragement and frank criticisms.

Finally, the Authors would like to thank Mario Girone, Carlo Baccigalupi, Liliana Viglione, Domenico Caiazza, Alessandro Parrella, Luca Sabato, Stefano Troisi, Giordana Severino, Donato De Paola, and Mario Kazazi, who helped in the proofreading of the chapters.

Authors' gratitude and appreciation go to Joel Stein and Millicent Treloar of Momentum Press, for their support, incitation, and patience during the preparation of this book. Their professionalism and helpfulness has considerably helped the Authors in overcoming all the difficulties arisen during book planning and writing.

Last but not least, Pasquale Arpaia thanks Marida Corso Arpaia for her patience during his long absence from home and, above all, for her impassable capability of supporting his work with a timeless and noble smile.

CONVENTION ABOUT THE NOTATION

Throughout the text of the book, we use the following notation common in software development.

All the names of components, subcomponents, superclasses, abstract classes, classes, and objects are written in italics and with the initial letter in capital.

Some technical terms at the first occurrence are reported in italics in order to emphasize their importance by definition.

All the names of well-known techniques or components/projects have the first letter in capital case.

Typical examples for the sake of reader ease are:

- a specific software component: the *Synchronizer* (italics and first letter capitalization)
- a specific class: the *FaultDetector* (italics and first letter capitalization)
- a well-known software technique: Object-Oriented Programming (first letter capitalization)
- a well-known software language: Measurement Domain Specific Language (first letter capitalization)
- a well-known software component: Graphical User Interface (first letter capitalization)
- a software project: Flexible Framework For Magnetic Measurement (first letter capitalization)
- a hardware component: Micro-Rotating Unit (first letter capitalization).

INTRODUCTION

First we build the tools, then they build us.

—Marshall McLuhan

In a modern test and measurement laboratory, one of the key activities in carrying out an assigned test is the realization of proper software for an automatic station. In small-size laboratories, test requirements vary over time, according to the requests of the market or the operating environment. This activity of test production turns out to be significantly expensive and troublesome. When a batch of objects to be tested arrives, and the test is too burdensome to be carried out manually, an automatic measurement station is developed. Test engineers define the measurement requirements that are used in designing and developing the automatic station. If the batches of objects under test are of small size and frequent occurrence, the activity of the bench automation predominates the test execution. Most significant efforts are devoted to instrumentation interfacing and software development.

In past years, the problem of easy-to-assemble and -configure hardware has been progressively faced encountered and effectively solved. Standard interfaces for operating the instrumentation remotely by PC (e.g., IEC 626, VXI, PXI, and so on) have become more and more widespread. Furthermore, devices based on automatic switches for cabling automatically the measurement circuit in order to carry out sequences of different tests on the same object under test have been successfully defined and made available on the market (e.g., ADLINK PXI-7921 [1] and National Instruments (NI) PXI-2529 [2]).

For the software, a different strategy has been followed, going through an approach of purely abridging the programming, such as for the standard de-facto LabVIEW™ by National Instruments [3]. A graphical programming language, the G, exploits graphical icons and wires symbolizing the data flow to make the program development easier. The approach is to conceptualize the application in terms of the involved objects and the exchanged data among them. However, the temporal sequence of the single actions to be executed in the measurement procedure is hidden in specific graphical constructs organized as frames in a movie. Conversely, imperative programming languages (e.g., C, Python, and so on) point out the operation's order and allow the temporal constraints of a measurement procedure to be managed in an easier way.

However, by this approach, for a laboratory operating with highly and rapidly varying test requirements, although the development effort is reduced indirectly by simplifying the programming phase, the software production quality, in terms of flexibility, usability, and maintainability, is not fostered intrinsically. An example of this problem is the experience gathered by the Authors at the European Organization for Nuclear Research (CERN), in the context of the magnetic measurements for the world's largest and highest-energy particle accelerator, the Large Hadron Collider (LHC) [4]. The LHC accelerates two counter-rotating particle beams with an energy of 7 TeV and forces them to collide at four intersection points. Magnetic fields up to 8.33 T are required to bend and focus the particle beams. A high current density is required to generate such a field level. For this reason, a superconducting magnet system was designed, including 1232 dipoles and 392 quadrupoles to bend and focus respectively the particle beams along their circular trajectories. The beam control imposes stringent constraints on the field quality, needing to be supported by adequate techniques for magnet testing [5, 6]. In the last few years, fast transducers (rotating units [7]) have been developed in order to increase the bandwidth of magnet quality tests by two orders of magnitude (from 10 to 100 Hz), by keeping a typical resolution of 10 ppm, simultaneously [7, 8]. The transducers produce voltage signals to be integrated in order to get the magnetic flux, according to Faraday's law (such as in rotating coils, fixed coils, stretched wire, and so on) [9–12], and are complemented by other techniques (such as Hall plates) [13]. A multipurpose numerical measurement instrument, the Fast Digital Integrator (FDI), has been, therefore, developed at CERN in cooperation with the University of Sannio, with the aim of reducing the flux acquisition time down to 4 μs while increasing the metrological performance [14]. The new integrator was conceived with the specific aim of being general-purpose, as much as possible, in order to become a sound basis for satisfying a wide range of magnetic measurement requirements over the years.

Furthermore, after the end of the LHC series tests, and on the medium term, the expectation is to have a number of very specific tests to be rapidly adapted and performed on single prototypes or relatively small batches of magnets [15]. These tests require the control of various devices, such as transducers, actuators, trigger and timing cards, power supplies, and other devices, not yet completely specified. Moreover, for different measurement techniques, different algorithms have to be implemented. All these conditions demand for re-engineering the measurement and acquisition software in order to be adequate for the new measurement requirements, and to manage the challenge of the new hardware.

In software development, a framework is a defined support structure for organizing and developing another software project [16]. It may include support programs, libraries, a scripting language, and other software tools to help develop and connect together the different components of a software project. Frameworks are designed mainly for simplifying software production, by allowing design engineers to spend more time on the application requirements, rather than on the low-level implementation details.

According to the Object-Oriented paradigm, a framework can be seen as a partial design and implementation for an application in a given domain [17]. It is described by a set of abstract classes, and instances of how those classes collaborate. The functionalities and architecture of the framework can be adapted and combined to create complete applications. Thus, frameworks allow a high level of reuse in Object-Oriented systems and a considerable reduction in the effort necessary for the realization of new applications.

While test programs are usually designed to solve specific problems with extremely limited capability to evolve, a framework for a given measurement domain, suitably conceived [17] in order to satisfy a wide range of requirements, could constitute a unified solution to drive all the existing and future park of measurement applications.

Namely, it is an effective way to implement a flexible measurement system, specifically useful when a test laboratory, in an industrial test division or a research center, is called to face highly and fast-varying test requirements [18–20]. A number of developments worldwide try to address this issue. Bosch applied the concept of framework in the measurement field, by proposing an Object-Oriented project capable of satisfying a wide range of applications [21]. Although this proposal leads to solving the issues of measurement software reusability and quality, the drawback of a higher programming skill for the test engineer arises.

In practice, the ideal situation would be to have a flexible software framework providing a robust library to control remotely all the instrumentation involved in the tests, including the new high-performance hardware, as well as all the tools the test engineer will need in the design of new measurement procedures.

A number of developments worldwide are trying to address these issues. At the commercial level, National Instruments (NI) proposes the product NI TestStand [22] for supporting the design of new test applications,by integrating software modules developed in different programming languages (C, C++, LabVIEW). However, NI TestStand does not support the development of single software modules, and, as a result, standard development and reusability are intrinsically limited. The Front-End Software Architecture (FESA) paradigm, adopted at CERN for the LHC controls [23] was developed to provide a suitable front-end for all the PCs interfacing with the LHC control instruments. However, the analysis of this software showed that a strong collaboration and involvement at the lowest level of FESA would be required in order to adapt the architecture to the aforementioned applications. At the Fermi National Accelerator Laboratory (FNAL), a new software system to test accelerator magnets was developed to handle various types of hardware, as well as to be extensible to all the measurement technologies and analysis algorithms [24]. Other subnuclear research centers (Alba, Soleil, Elettra, and ESRF) collaborated in order to develop a suitable software framework for testing accelerator magnets [25]. This Consortium proposes TANGO, an Object-Oriented system to handle different measurement applications. At that time, the software of FNAL and the Object-Oriented system Tango were still under development and not yet accessible worldwide.

The exposition of this book covers the specification, design, implementation, and validation of a software framework for measurement control and data acquisition. The framework's objective is it to maximize the measurement software quality, in terms of flexibility, reusability, maintainability, and portability, by simultaneously keeping high efficiency levels. Moreover, the framework can be configured for satisfying a large set of measurement applications in a specific measurement application field. It is characterized by (a) flexibility in the rapid and cost effective realization of "scriptable" applications, including prototyping in an R&D context, (b) a modular architecture to mix and reuse components chosen from an incremental library, and (c) high performance to exploit the increased throughput of the new transducers and acquisition systems.

The examples of framework applications in this book refer to an implementation in C++, mainly based (a) on *Object-Oriented Programming* (OOP) and (b) on an innovative technology, the *Aspect-Oriented Programming* (AOP) [26]. The latter technique extends the objects capabilities in order to encapsulate features transversal to several functional units (*crosscutting concerns*) by means of new software modules, the *aspects*.

Moreover, after the kernel and the main components of the framework are available, the framework's key user, the test engineer, needs an easy and fast way to formalize the test procedure by means of measurement scripts. To achieve this goal, a *Measurement Domain Specific Language* (MDSL) is presented in this book.

Finally, also suitable means for automatically generating user interfaces are illustrated. The practical goal is to allow programmers, as test engineers, who are not typically trained to design graphical interfaces, to easily produce good Graphical User Interfaces (GUIs) for their applications.

In the following, the main contents of the book are outlined.

In Chapter 1, an overview of the world of software for measurement and test applications is given. For this purpose, the application software is classified in two main groups: (a) measurement software in market solutions and (b) software in the state of the art of research. In the first group, the main solutions of measurement software producers are reviewed, by establishing criteria for classifying the related products. In the second group, software applications developed in the field of instrumentation and measurement research are examined, by looking for a conceptual classification of the related scenario.

In Chapter 2, software frameworks are presented and the main related concepts are introduced. Subsequently, these principles are examined in the more specific case of software frameworks for measurement automation applications, by discussing their rationale and main features. Then, a section is dedicated to the introduction of Domain Specific Languages (DSLs). Finally, the requirements and desirable features of a software framework for measurement applications are presented in relation to the different types of users who would interact with such a system.

Chapter 3 continues the travel throughout the software for measurement applications, by focusing attention on one of the most important paradigms of the last decades: the Object-Oriented Programming. First, the main concepts of object-oriented programming are presented. Then, main concepts of design patterns of Object-Oriented Programming are highlighted. Finally, the main concepts of Aspect-Oriented paradigm are illustrated by emphasizing the advantage of its use in measurement applications.

Chapter 4 presents the design of a software framework for automatic measurements application, based on Object-Oriented and Aspect-Oriented programming. Initially, the paradigm of the framework is introduced, by highlighting the basic ideas leading to its conception and design, as well as its architecture at the structural and functional level. Then, the main components of the framework are described, by firstly introducing the corresponding state of the art, the leading concepts, and the architecture with their main modules. The review starts with the *Fault Detector*, aimed at identifying and locating failures and faults, transparently, to the user. Fault detection is a crosscutting concern; therefore, the design is led by an Aspect-Oriented approach. Then, the *Synchronizer*, aimed at coordinating measurement tasks with well-defined high-level software events (e.g., start and stop, or device events), is presented. The review continues with the Measurement Domain Language, aimed at specifying complete, easy-to-understand, -reuse, and -maintain applications efficiently and quickly by means of a script. Finally, the Automatic Generator of User Interface, aimed at separating the user interfaces easily from the application logic for enhancing the flexibility and reusability of the software, is illustrated.

In Chapter 5, the assessment of the software quality for measurement software frameworks is presented. First, main concepts of software quality are introduced very synthetically from a general perspective. Then, the approach proposed in the standard ISO 9126 is chosen as reference model for the quality assessment of measurement and test software. Finally, a method based on specific metrics for assessing the degree of flexibility achieved by a software framework for measurement applications is presented.

In Chapter 6, an overview of a case study of the software framework for magnetic measurements realized at CERN in cooperation with the University of Sannio, the FFMM [27], is provided. Initially, a background on the application context of testing magnets for particle accelerators is given. In the second part of the chapter, an outline of the FFMM project is given. In particular, the FFMM is presented by introducing its main design concepts and architecture. Subsequently, its main components, the *Fault Detector*, and the *Synchronizer*, are described, by highlighting their architectures and the implementation of their classes. Finally, the application of the Measurement Domain Specific Language (MDSL) and the Graphical User Interface engine are reported.

In Chapter 7, some implementation examples in C++ code of the most significant part of the framework FFMM at CERN are illustrated. The examples are chosen by referring to the main layers of the framework architecture. In particular, for each service layer, implementation details about the layer structure and some devices classes are shown.

In Chapter 8, the validation of the framework components, the *Fault Detector*, the *Synchronizer*, the MDSL and the *Advanced Generator of User Interfaces* is treated. Each component was validated in different case studies. The *Fault Detector* was analysed in a rotating coil system for magnetic measurements; the *Synchronizer* in a magnetic permeability measurement system; the MDSL in a superconducting magnet testing and magnetic permeability measurement, and the *Advanced Generator of User Interfaces* in a magnetic permeability measurement system.

In Chapter 9, a specific discussion on the goal achievement in functional terms is presented for the FFMM as a whole. FFMM was used to develop applications for different measurement activities currently carried out at CERN. Different application scenarios are a good test bed for checking FFMM's capability on offering an environment for a fast development of several measurement applications with different requirements. The discussion is focused on describing the measurement procedure, the test station, and the experimental results for measuring the magnetic permeability by means of the split-coil permeameter and the magnetic field by means of the rotating coils, as well as for testing and compensating the field distortion of the superconducting cryo-magnets of the LHC. Finally, a specific assessment of the flexibility of the FFMM is presented on the basis of the method presented in Chapter 5. The impact of adding or modifying a device, changing service strategies, and implementing new measurement algorithms is highlighted.

REFERENCES

[1] "PXI-7921: 24-CH 2-Wire Multiplexer Module." ADLink Technologies: Switches. http://www.adlinktech.com/PD/marketing/Datasheet/PXI-7921/PXI-7921_Datasheet_1.pdf

[2] "High-Density Multiconfiguration Matrix Modules: NI PXI-2529 Specifications." National Instruments. http://sine.ni.com/ds/app/doc/p/id/ds-375/lang/en.

[3] National Instruments. n.d. What is NI LabVIEW? http://www.ni.com/webcast/2696/en/

[4] CERN. 2004. LHC Design Report. CERN 2004-003.

[5] Bottura, L., and K.N. Henrichsen. September 2004. "Field measurements." CERN Laboratory for Particle Physics. http://cds.cern.ch/record/597621/files/lhc-2002-020.pdf

[6] Amet, S., L. Bottura, L. Deniau, and L. Walckiers. March 2002. "The Multipoles Factory: An Element of the LHC Control." *Applied Superconductivity, IEEE Transactions* 12, no. 1, pp. 1417–21. doi: http://dx.doi.org/10.1109/tasc.2002.1018668

[7] Brooks, N.R., L. Bottura, J.G. Perez, O. Dunkel, and L. Walckiers. June 2008. "Estimation of Mechanical Vibrations of the LHC Fast Magnetic Measurement System." *Applied Superconductivity, IEEE Transactions* 18, no. 2, pp. 1617–20. doi: http://dx.doi.org/10.1109/tasc.2008.921296

[8] Haverkamp, M., L. Bottura, E. Benedico, S. Sanfilippo, B. ten Haken, and H.H.J. ten Kate. "Field Decay and Snapback Measurements Using a Fast Hall Plate Detector." *Applied Superconductivity, IEEE Transactions* 12, no. 1, pp. 86–89. doi: http://dx.doi.org/10.1109/tasc.2002.1018357

[9] Elmore, W.C., and M.W. Garrett. 1954. "Measurement of Two-Dimensional Fields, Part I: Theory." *Review of Scientific Instrument* 25, no. 5, pp. 480–85. doi: http://dx.doi.org/10.1063/1.1771105

[10] Dayton, I.E., F.C. Shoemaker, and R.F. Mozley. 1954. "Measurement of Two-Dimensional Fields, Part II: Study of a Quadrupole Magnet." *Review of Scientific Instruments* 25, no. 5, pp. 485–89. doi: http://dx.doi.org/10.1063/1.1771107

[11] DiMarco, J., and J. Krzywinsky. March 1996. "MTF Single Stretched Wire." Technical Report, Fermi National Accelerator Laboratory.

[12] DiMarco, L., H. Glass, M.J. Lamm, P. Schlabach, C. Sylvester, J.C. Tompkins, and J. Krzywinsky. 2000. "Field Alignment in Quadrupole Magnets for the LHC Interaction Region." *IEEE Transactions on Applied Superconductivity* 10, no. 1, pp. 127–30. doi: http://dx.doi.org/10.1109/77.828515

[13] Bottura, L., L. Larsson, S. Schloss, M. Schneider, and N. Smirnov. 2000. "A Fast Sextupole Probe for Snapback Measurement in the LHC Dipoles." *IEEE Transactions on Applied Superconductivity* 10, no. 1, pp. 1435–38. doi: http://dx.doi.org/10.1109/77.828509.

[14] Arpaia, P., V. Inglese, and G. Spiezia. July 2009. "Performance Improvement of a DSP-Based Digital Integrator for Magnetic Measurements at CERN." *Instrumentation and Measurement, IEEE Transactions* 58, no. 7, pp. 2132–38. doi: http://dx.doi.org/10.1109/tim.2008.2006723

[15] Arpaia, P., L. Bottura, M. Buzio, D. Della Ratta, L. Deniau, V. Inglese, G. Spiezia, S. Tiso, and L. Walckiers. May 1–3, 2007. "A Software Framework for Magnetic Measurements at CERN." *Proceedings of the Instrumentation and Measurement Technology Conference.* Warsaw, Poland: IEEE.

[16] Anderson, J.L. 2005. "How to Produce Better Quality Test Software." *IEEE Instrumentation and Measurement* 8, no. 3, pp. 34–38. doi: http://dx.doi.org/10.1109/mim.2005.1502445

[17] van Gurp, J., and J. Bosch. 2001. "Design, Implementation and Evolution of Object-Oriented Frameworks: Concepts and Guidelines." *Software Practice and Experience* 31, pp. 277–300. doi: http://dx.doi.org/10.1002/spe.366

[18] Hashempour, H., F. Lombardi, W. Necoechea, R. Mehta, and T. Alton. 2007. "An Integrated Environment for Design Verification of ATE Systems." *IEEE Transactions on Instrumentation and Measurement* 56, no. 5, pp. 1734–43. doi: http://dx.doi.org/10.1109/tim.2007.895611

[19] Guo, Z., P. Chen, Y. Feng, Y. Jiang, and F. Hong. 2010. "ISDP: Interactive Software Development Platform for Household Appliance Testing Industry." *IEEE Transactions on Instrumentation and Measurement* 59, no. 5, pp. 1439–52. doi: http://dx.doi.org/10.1109/tim.2010.2040931

[20] Anderson, J.L., R. Rajsuman, N. Masuda, and K. Yamashita. 2005. "Architecture and Design of an Open ATE to Incubate the Development of Third-Party Instruments." *IEEE Transactions on Instrumentation and Measurement* 54, no. 5, pp. 1678–98. doi: http://dx.doi.org/10.1109/tim.2005.856714

[21] Fayad, M.E., and R.E. Johnson. 1999. *Domain-Specific Application Frameworks*, pp. 177–205. New York, NY: Wiley.

[22] "TestStand." n.d. National Instruments. http://www.ni.com/teststand/

[23] Guerrero, A., J.J. Gras, J. Nougaret, M. Ludwig, M. Arruat, and S. Jackson. 13–17 October 2003. "CERN Front-End Software Architecture for Accelerator Controls." 9th International Conference on Accelerator and Large Experimental Physics Control Systems, Gyeongiu, Korea, pp. 342–44.

[24] Nogiec, J.M., J. DiMarco, S. Kotelnikov, K. Trombly-Freytag, D. Walbridge, and M. Tartaglia. 2006. "A Configurable Component-Based Software System for Magnetic Field Measurements." *IEEE Transactions on Applications Superconducting* 16, no. 2, pp. 1382–85. doi: http://dx.doi.org/10.1109/tasc.2005.869672

[25] Abeille, G., S. Pierre-Joseph, J. Guyot, and M. Ounsy. 10–14 October 2011. "Tango Archiving Service Status." In *13th International Conference on Accelerator and Large Experimental Physics Control Systems*, pp. 127–30. Grenoble, France: EFDA.

[26] Arpaia, P., M. Bernardi, G. Di Lucca, V. Inglese, and G. Spiezia. 2010. "An Aspect-Oriented Programming-based Approach to Software Development for Fault Detection In Measurement Systems." *Computer Standards & Interfaces* 32, no. 3, pp. 141–52. doi: http://dx.doi.org/10.1016/j.csi.2009.11.009

[27] Inglese, V. 2009. A Flexible Framework for Magnetic Measurements (CERN-THESIS-2010-019, CERN).

PART I

Background

CHAPTER 1

SOFTWARE FOR MEASUREMENT APPLICATIONS

The real danger is not that computers will begin to think like men,
but that men will begin to think like computers.

—Sydney J. Harris

1.1 OVERVIEW

Main objective of this chapter is to analyze the world of software for measurement and test applications. With this aim, two main groups are considered: (a) measurement software in market solutions and (b) software in the state of the art of research. In the first group, the main solutions of producers are reviewed, by establishing criteria for classifying the related products. In the second group, software applications developed in the field of instrumentation and measurement research are examined, by looking for a trend in the related scenario.

1.2 BASICS

Automatic test and measurement (or data acquisition) systems are used to assess experimentally parameter values of a process or a product. They are different from the monitoring systems integrated in process/plant supervision and control, because the measured values are not used directly (i.e., as a part of the same system) to automatically adjust the system under test [1].

Moreover, for both the cases, increasing automation in the production process amplifies more and more the need for accurate measurement systems for quality control and, consequently, the need for developing related software. The choice of the right application software for an automatic measurement system is an important step for all the test operators. As a matter of fact, the software is the core of modern automatic measurement systems, and the main imperative in the selection of a development tool is the scalability as the measurement systems matures, in order to avoid rewriting the code.

But software developers also face the challenge of producing reliable and high-quality applications in a very-short time [1]. Moreover, another ongoing need is the integration of different families of various measurement instruments, by minimizing configuration and measurement time. Finally, the software for measurement applications should be highly effective in exploiting first the increased throughput of the new transducer and acquisition systems and, at the same time, keep flexibility, reusability, maintainability, and portability, in order, to maximize its quality. In particular, in this context,

1. *flexibility* means the capability of satisfying a large set of measurement applications, possibly in different fields, with rapid and cost-effective realization, especially in research and development environment;
2. *reusability* of components is aimed at adding new features with minor code alterations, as well as with reduced implementation time and bugs probability, owing to the preceding testing and use;
3. *maintainability* is aimed both at an easy revision after release, in order to correct faults and improve performance or other attributes and error correction, and at optimizing performance and features, by removing obsolete capabilities; in fact, during the software life, evolution is unavoidable, thus suitable mechanisms must be set up for crafting, assessing, and tracing modifications; and
4. *portability*, by suitably abstracting between application logic and system interfaces, for exploiting the same code in diverse environments, in order to reduce development cost when producing for several computing platforms.

In the upcoming sections, main measurement software applications are reviewed in order to highlight key features, analyze principal market solutions, companies, and products (Section 1.3). Then, the horizon of the investigation is broadened to the software for research and development, in order to focus on the main motivations and needs in current innovation trends (Section 1.4).

1.3   MAIN MARKET SOLUTIONS

In this section, first the main criteria for addressing the right choice of software for automatic measurement systems are highlighted; then, the corresponding main market solutions are analyzed.

1.3.1   CRITERIA FOR CHOOSING SOFTWARE

Generally, the main criteria for choosing the right software environment for developing successfully and effectively a high-quality measurement application are the *flexibility* respect to the application, the *usability* for nonspecialized users, the *integrability* with existing applications, the *information availability*, and the software *history* (i.e., stability and longevity) [2].

About *flexibility*, software applications range from ready-to-run programs to environments for developing fully customizable applications. Although the choice of a software environment is based on the current development requirements, an important factor is how the application can scale and solve problems over time. Specific measurement procedures use software tools with a set of features for a limited subset of hardware options. The current and over-time scaled requirements are met by suitably selecting a development environment allowing custom applications to be created. Development environments for measurement application are extremely flexible, allowing instrumentation drivers to be integrated into the software and a custom user interface to be developed. The only trade-off is the time spent learning their use and the related programming language. Although this could seem a negative aspect, the development environments provide a variety of information resources, including online and live training, getting-started examples, community forums to share code, and so on.

Another key criterion for a good selection is the *usability* of the software application, that is, the time needed to learn the new software. This is different for each user and depends on the type of the software application. Users spend considerable time to learn the language used within the development environment. Ready-to-run software tools are the easiest and fastest to learn, because programming details are excluded from the users. If the choice of a new environment requires one to learn a new programming language, the users consider the software as aimed at solving the current application problem, rather than at low-level encoding. Programming languages, such as C++, are more challenging to learn for their complex grammar and syntax rules. Graphical programming languages are easier to learn for their intuitive nature, but are not as widespread as standard de facto traditional languages [2]. Moreover, their level of abstraction is higher, and does not provide the possibility of programming at low-level by accessing directly hardware features for optimizing time and memory use in constraining real-time applications.

Integrability is the capability of the application software to be interfaced effectively with instrumentation and measurement setup, and with other software tools for data analysis, visualization, and storage. In most cases, the existence of a software driver of a measuring device is not sufficient for its direct integration into the automatic measurement system. The most important thing is to choose a software driver or a tool directly compatible with the software application as a whole. In most measurement applications, tools for data analysis are designed directly for data acquisition through suitable signal manipulation aids [2]. Furthermore, the application software usually should have an easy way to visualize the acquired data. Finally, the application software should integrate easily with system and data management software to store data of tests.

About *information availability*, the chosen development software should be surrounded by a certain amount of knowledge. A wealth of resources and information makes learning a new software tool easy for each user, by guiding the application development with suitable feedback. Before choosing a development tool, the best practice is to take the information about existing case studies and gain as much knowledge as possible about their positive and negative aspects.

The *reuse* and *scalability* of a measurement system depends on the stability and longevity of the application software, in order to avoid obsolescence in short time. For software *reuse*, the abstraction plays a central role in order to make intuitive and expressive the application software. Moreover, *scalability*, conceived as the "ability of a program to adapt effectively to variation in data size," allows the measurement software to process well both small and large data sets, independently of the device setup size.

1.3.2 MARKET LEADERS AND PRODUCTS

In this subsection, the software market for measurement applications is analyzed according to the aforementioned criteria in order to depict a scenario of the commercial products. The resulting tree of the main market solutions is represented in Figure 1.1, highlighting producers and their software. A further important objective criterion is that the producers considered in Figure 1.1 have been leaders in the market for at least the last 30 years. The tree does not have the presumption to be neither exhaustive

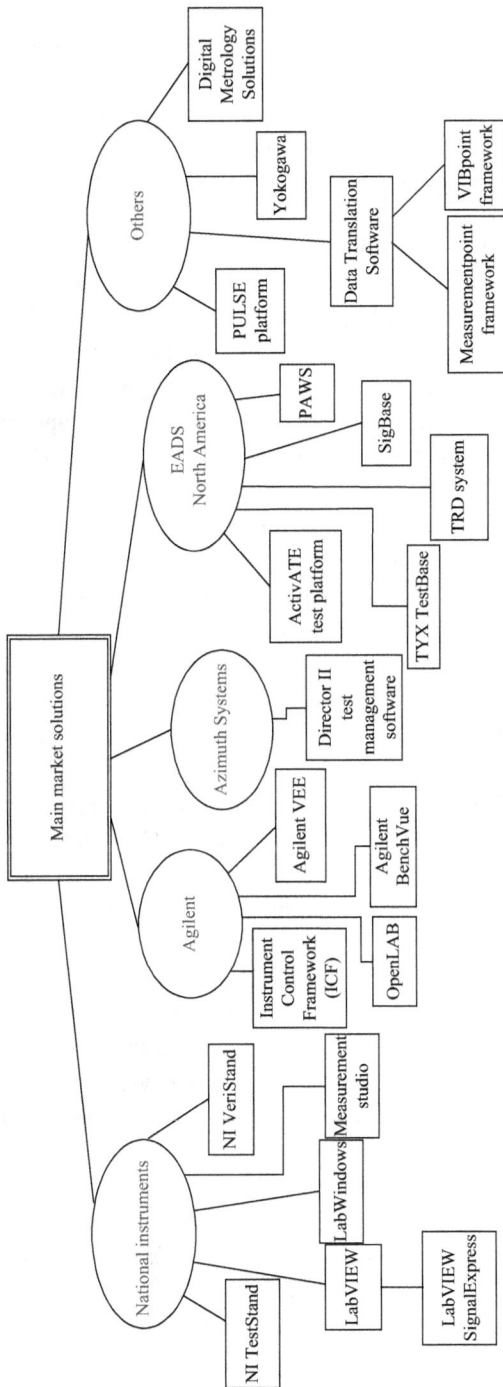

Figure 1.1. Main market solutions tree: Leaders and products.

nor objective, in the sense that the classification is made according only to the earlier criteria and, above all, for purposes functional to the book exposition. Moreover, the classification does not express any judgment about a scale of quality of the products, and reflects only personal ideas of the Authors. Finally, the corresponding market picture is taken at the moment of the book publication, without any presumption to be durable over a long time.

In particular, the following main producers are highlighted: (1) National Instruments, (2) Agilent Technologies, (3) Azimuth Systems, (4) EADS North America, and (5) others. Indeed, these producers are the leaders of measurement application systems, as producers, above all, of measurement instrumentation (devices, sensors, transducers, analyzers, and so on), and, only after, of software for measurement application. This aspect is common to all the producers, main leaders, and other producers (Figure 1.1).

By analyzing the products, the flexibility is not really a global factor, but in some case it depends on the application field. Software integrability is subject to the use of proprietary instrumentation. Regard to usability, resource availability, and history, all the producers are leaders in the market for at least 30 years, and for this reason their products are recognized as stable and reliable.

In the following, a short overview of the producers and their main products is presented.

National Instruments (NI) offers a complete software and hardware platform for building an automatic test system [3]. By means of this platform, NI offers service programs for supporting users throughout the application life cycle—from planning and development up to deployment and on-going maintenance. According to the aforementioned basic criteria, the highest performance NI products are the Test Management and Development Software (NI TestStand, NI VeriStand, NI LabVIEW, NI LabWindows/CVI, and NI Measurement Studio) [3]. NI TestStand is a test management software for speeding the development of automated test and experimental validation systems without programming knowledge. Test sequences integrating code modules written in any test programming language can be developed. The sequences specify execution flow, database logging, connectivity, and reporting to other enterprise systems. NI VeriStand allows real-time tasks (I/O, profiles, and alarms) to be configured, system simulations to be implemented, and a run-time editable user interface to be created. Control algorithms, simulation models, and other tasks from NI and third-party software environments can be imported.

LabVIEW [4] is an intuitive graphical programming environment to help the user in quickly developing complex test software with support for thousands of technologies and instruments, such as multicore and field-programmable gate arrays (FPGAs). LabVIEW platform has seen its user base grow to more than one million consumers worldwide this past decade, to become a true de facto standard. The programming language of LabVIEW is a dataflow programming language, inspired by G-code. The execution structure is determined by a graphical block diagram (different function-nodes connected by drawing wires). In addition to graphical programming, the main benefits of LabVIEW are: (a) interfacing, as access to many different drivers for devices and hardware instrumentation; (b) large libraries, a large number of function block for data acquisition, analysis, mathematic, signal processing, and so on; (c) code reuse, reusability of code without modification; and (d) parallel and multithreading programming, easy to program multitask for hardware automation.

In the LabVIEW platform, further measurement application software can be found, such as **LabVIEW SignalExpress**. SignalExpress is an interactive software package for data-logging, as well as for acquiring and presenting data from hundreds of data acquisition devices and instruments, by allowing their analysis without programming. From a technical point of view, LabVIEW SignalExpress is a signal analysis software package, because the core of its features is specifically customized for this task.

LabWindows/CVI is a proven ANSI C integrated development environment, providing test engineers with a comprehensive set of programming tools, including instrument drivers and functions, analysis libraries, multicore programming capabilities, and assistants to auto-generate code. With respect to LabVIEW, hardware features of the instruments can be accessed at low level owing to the programming in C.

Measurement Studio is a plug-in of NI for Microsoft Visual Studio [5], considerably reducing application development time for creating test applications, by providing an integrated suite of measurement and automation controls, tools, and class libraries, specific for .NET programmers. The developer productivity is increased by extending the Microsoft .NET Framework and by providing measurement and automation classes as well as Windows Forms and Web Forms controls for Visual Basic .NET and Visual C#. Most important features of Measurement Studio are the advanced analysis and signal processing libraries, the engineering-specific user interface controls, the hardware integration, the compatibility with NI TestStand, the automation of the test process, and the flexible debugging.

Agilent Technologies represents a company leader in the field of measurement instrumentation and application software [6]. Its software and informatics portfolio covers a broad range of analytical workstations, applications, workflows, and laboratory management solutions to meet the needs of the life sciences and chemical industries. The analysis of Agilent's software and informatics portfolio points out many software tools for measurement application, but only few of them meet the criteria of the present analysis. Particularly interesting is the **"Liquid Chromatography: Instrument Control Framework (ICF)"** [7]. The ICF is a software package that makes fast and easy for third-party software providers the control of Agilent liquid chromatography systems through their data systems or workstations. ICF contains instrument drivers and user interfaces, and provides a simple programming interface for third-party software connectivity. The main technical feature of ICF is the application-oriented nature of software, though limited to liquid chromatography systems, and the strict linking with specific Agilent instrumentation. **OpenLAB Laboratory Software Framework** is an Agilent software tool with a new approach for managing laboratory data [8]. Agilent OpenLAB is more than a laboratory database, in particular it represents a framework for managing laboratory content and work processes. The main features of OpenLAB are: scalable architecture, protect and secure data, centralized management, support for laboratory business processes, and flexible deployment [8]. **Agilent VEE (Visual Engineering Environment)** [9] is a graphical language environment, designed to provide a quick path to measurement and analysis. The main features of VEE are: (a) a virtual environment (objects or instruments, wires, data flows, and so on), to design the measurement setup; (b) tools to create and to debug programs easily; and (c) multithreading technology and multicore programming. **Agilent BenchVue** [10] is an intuitive software environment that accelerates testing by providing multiple instrument measurement visibility and data capture without programming. The main specifications of BenchVue are: viewing, capturing, and exporting measurement data.

Azimuth Systems is a leading provider of wireless broadband test equipment and channel emulators for LTE-Advanced, LTE, WiMAX, and other 2G/3G/4G technologies [11]. Azimuth's products are used by the world's foremost wireless semiconductor and system vendors, as well as by leading service providers, to speed time-to-market and improve wireless service quality. Azimuth's fully integrated software architecture is designed to efficiently manage all the aspects of wireless testing, from test configuration to results analysis. In particular, **Azimuth DIRECTOR™ II** is a flexible software management tool that enables automation of custom test plans, runs standard scripts, and manages all the devices in the testbed [12]. Director II has a modular architecture, with five primary applications: (a) *TestBed Manager*, providing a common, centralized interface to control and manage all the test bed devices within the test environment; (b) *Test Builder*, enabling engineers to easily construct flexible test scripts without knowledge of Test Control Language programming; (c) *Test Driver*, providing an intuitive interface to Azimuth's standard scripts for performance and certification and to custom developed scripts; (d) *Test Scheduler*, a batching and scheduling application allowing users to organize and batch a group of test scripts, with sequence and scheduling flexibility; and (e) *Test Editor*, a script editor enabling engineers to create their own customized test scripts.

EADS North America Test and Services, a division of EADS North America, is a complete test and measurement systems provider [13], offering a full-range of products and capabilities, such as integrated test solutions, complete turn-key hardware and software systems, custom designs, commercial-off-the-shelf (COTS) instrumentation, engineering solutions, and test software. EADS North America Test and Services has a leading role in software development for test instrumentation that supports flight hardware, guidance systems, aircraft engines, and other mission-critical systems. The **ActivATE** platform and the **TYX** product line (**PAWS**, **TestBase**, **SigBase**, and Test Requirements Document (**TRD**), not pointed out in Figure 1.1 for conciseness) represent the main software products [14] in the measurement application software. ActivATE is a platform to develop software, providing both runtime and development environments for Automatic Test Equipment (ATE). It is an open architecture Integrated Development Environment (IDE), allowing the user to rapidly and intuitively build test programs for ATE. The ActivATE™ Test Platform is a software environment based on the .Net architecture. **PAWS** Developers Studio is a product to compile, modify, debug, document, and simulate the operation of the Abbreviated Test Language for All Systems (ATLAS), a military standard language for automatic testing of avionics equipment [15] test programs in a modern Windows NT Platform environment. A full range of the most-commonly used ATLAS Language subsets is supported. A PAWS Toolkit can modify the ATLAS Language subset to meet the particular ATE configuration. **TestBase** is a test executive supporting the visual development, database storage, and run-time execution of test strategies (also known as test plans or test sequences). TestBase modular and open architecture enables system integrators and end-users to customize and extend the product and to integrate additional third-party applications. The entire ATE lifecycle (new, rehost, and legacy) is supported through common user interfaces, configuration management capabilities, and automatic generation of development documentation. **SigBase** is an

IEEE-1641 [16] compliant signal-based test environment, supporting the visual organization and execution of IEEE-1641 strategies created from basic signal components and test signal frameworks. SigBase is also an International Traffic in Arms Regulations (ITAR)-controlled product [17], that is, SigBase is considered a defense-related service, so the export and import of this is controlled by ITAR.

The **TRD** System helps users to develop and document the strategy and structure of test programs, using suitable screens, flowchart generation tools, and documentation formats, all in accordance with the most popular military standard formats or with unique custom formats. ATLAS test program can be automatically generated from the information provided by the user. Like SigBase, TRD is an ITAR-controlled product.

The producers labeled *others* (Figure 1.1) represent a group of measurement application creators of market impact not comparable to the aforementioned companies. Indeed, **PULSE platform**, **Data Translation**, **Yokogawa**, and **Digital Metrology Solutions** have products meeting the previously mentioned basic criteria of flexibility and integration of software. Most important drawbacks are related to the resource availability, usability, stability, and general history. This makes it impossible as a plausible comparison with the main leader producers.

From a strict technical point of view, all the aforementioned products of all the producers can be classified in four categories: **Test Management Software**, **Test Development Software**, **Application-based Software**, and **Signal Analysis Software**. A corresponding cataloging of the technical categories versus producers of software packages is plotted in Figure 1.2.

The Test Management Software products (Director II, ActivATE Test Platform, NI TestStand, and VeriStand) are flexible measurement software packages integrated in a full management tool. They foster and manage the most important measurement steps: (1) management of all the devices within the test environment, (2) flexible construction of test scripts, (3) test driver by an intuitive interface, (4) test batching and scheduling, and (5) test editor for implementing measurement scripts. The main difference between the classes of Test Management Software and Test Development Software products is the management aspect. In effect, test development software products are measurement–oriented, that is, they represent suitable environments for developing a measurement application. Some products (Instrument Control Framework, PAWS, TRD System, Pulse Platform, and VIBpoint) show management and development features, but in specific measurement step or application. As an example, ICF and Pulse Platform are full measurement application platforms, but their fields of application are specific, liquid chromatography and sound and vibration measurements,

Categories \ Producers	National Instruments	Agilent	Azimuth Systems	EADS	Others
Test Management Software	NI TestStand, NI VeriStand	OpenLAB	Director II	ActivATE test platform	
Test Development Software	LabVIEW, LabWindows, Meas Studio	Visual Engineering Environment		TestBase	MeasurementPoint Software
Application Based Software		Instrument Control Framework		PAWS, TRD systems	Pulse Platform, VIBpoint, Yokogawa Software
Signal Analysis Software	LabVIEW SignalExpress	BenchVue		SigBase	Digital Metrology Solutions Soft.

Figure 1.2. Plot of Technical Product Categories versus Producers.

respectively. For this reason, this kind of products difficultly matches other fields of application, and they are inserted in the category of Application-based Software. In the category of Signal Analysis Software, products such as LabVIEW Signal Express, BenchVue, and SigBase of EADS, for analyzing the signals in an easy and quick way are included.

1.4 RESEARCH: STATE OF THE ART

During the last decade, the need for automated tools specifically supporting quality control inside the actual production process is rising more and more. As a matter of fact, the increased quality and the improved automation level inside the production process require monitoring and control systems with higher and higher performance. This trend has dramatically enlarged the necessity for research and development of automatic measurement systems.

Although a substantial part of these systems is developed at a hardware level, that is, connected directly to the physical world through several sensors and actuators, an automatic measurement system contains a considerable amount of software. Furthermore, nowadays the need for an automatic support to their development is evident also [1]. As a direct consequence, this interest spurred the research in the field of software applied to measurement and test systems. The objective of this section is to analyze the research and the development of software in measurement applications, such as for industry, science, and other fields.

With respect to the previous section on main market solutions, the basic criteria for analyzing the state of the art in research on innovative software applications for automatic measurement systems are conceptually different. Talking about resource availability, history and stability, as well as usability is meaningless for scanning the research trends. Flexibility and integration concepts are again used, but mainly at a technical level. In scientific literature, the measurement-oriented software is integrated in automatic test or real-time monitoring systems. The software applied to measurement methods can be flexible with respect to the specific application field.

A more useful strategy for classifying the research state of the art is to analyze the software applied to automatic measurements from a technical point of view. From this viewpoint, in research papers, four main classes can be identified (Figure 1.3): *Hardware and Software Platforms*, *Specific and Custom Software*, *Application Field Software*, and *Development Environments*.

A *platform* includes a hardware architecture and a software framework combined to run application software. Typical platforms include operating

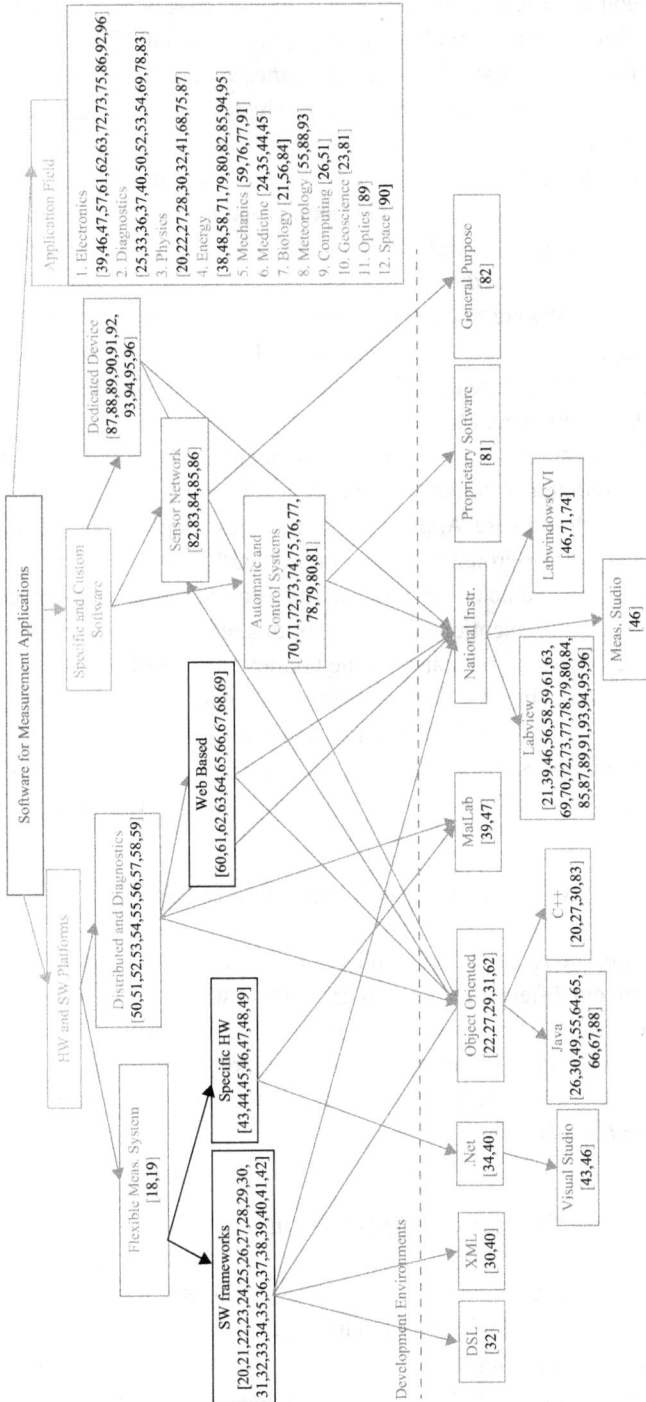

Figure 1.3. State of art Research Tree: Software for Measurement Applications.

system, computer architecture, programming languages, and related user interface (run-time system libraries or graphical user interface).

On the market, *specific or custom software* (also known as *bespoke* software or *tailor-made* software) is expressly developed for some definite organizations or users. In scientific research, the concept is analogous, but the organization and the user are replaced by specific instrumentation, devices, sensors networks ,or particular setup, applied and found in literature.

The third possible synthesis approach is to classify the software of research according to *the application field*, although the experimental measurement software touches multiple field of research.

Finally, the software can be classified (Figure 1.3) according to the *environment* or language used for implementing the measurement system (.Net, Object-Oriented, Matlab, etc.).

A development environment is a crucial element in software production. An environment might be defined in the simplest case, even, as a place to launch software. Application software can depend on the features of the particular environment, either on the hardware and operating system, or the virtual machine it runs on. This is a transversal criterion common to all the aforementioned classes, for this reason the related research trends were classified separately at the bottom of Figure 1.3.

In the following, the synthesis tree of Figure 1.3 is discussed, by referring to its previously mentioned main classes of research trends: (a) *Hardware and Software Platforms*, (b) *Specific and Custom Software*, (c) *Application Field Software*, and (d) *Development Environments*.

1.4.1 HARDWARE AND SOFTWARE PLATFORMS

In literature, the research trends related to *Hardware and Software Platforms* point out different features, mainly of hardware and software nature, or specific technical keywords, such as flexible, frameworks, distributed, and so on. Accordingly, the corresponding branch of the tree in Figure 1.3 is divided in two main trend sets: *Flexible Measurement Systems* and *Distributed and Diagnostics Systems*.

1.4.1.1 Flexible Measurement Systems

In general, research trends on flexible measurement systems put main design focus on flexibility, modularity, generality, and hardware independence [18]. They represent software architectures meeting specifically these requirements by innovative solutions, and structures expressly conceived for implementing versatile measurement systems. Typically, flexible measurement systems are based on software packages (monitoring with data acquisition, processing, transmission, and storing, as well as result analysis), with various possibility of communication [19]. They are applied to different industrial operation, such as the protection of plants operation, system supervision, fault detection, and provision of measured values. In research trends about flexible systems, the other two specific classes can be distinguished as: *Software Frameworks* and *Specific Hardware* systems emphasizing software and hardware aspects, respectively.

Software Frameworks

All the innovative measurement systems exploiting a software framework privilege flexibility and integration. A framework allows software for measurement and test applications under highly and fast-varying requirements to be developed easily. The framework can be configured in satisfying a large set of measurement applications in a generic field for an industrial test division, a test laboratory, or a research center [20]. A software platform can be designed for particular hardware or functional requirements, but with the specific aim of providing the development of new features [21, 22]. Other types of frameworks are general tools for automatically analyzing the measurement data [23–25], or the dependability of measurements and algorithms on distributed systems [26].

The generic Object-Oriented based frameworks present the main advantages of Object-Oriented Programming: encapsulation, inheritance, flexible-construction, and multitask options. These features make the measurement approach strongly oriented to software integration, with the effect of supporting dynamic modeling and data fusion of instruments. This allows both Object-Oriented databases to be created for measurement, and expert systems or test platforms to be built for processing measurement information [27–31].

A model-driven paradigm for defining measurement/test procedures, configuring instruments, and synchronizing tasks is applied to a flexible framework for maximizing the software quality [20, 32]. The final goal of the project, presented in [33], is to design, implement, and deploy an extensible, flexible, and dynamic system. A framework can be provided by a powerful mechanism to adapt component-based distributed applications to changing environmental conditions [34].

An increasing research trend is arisen aimed at conceiving measurement frameworks [35, 36] for general-purposes and rapid prototyping of applications for end-to-end sensing systems, in order to monitor and diagnose problems on a wide-area network [37]. In [38, 39], global framework for monitoring, control, and optimization of power electronics systems are described. Component-based frameworks for data stream processing are presented for highly distributed measurement systems [40, 41]. Other software development environment, framework-based, provide a graphically programming way to quickly build automatic test and measurement systems [42].

Specific Hardware

Sometimes, software packages are present in specific hardware-based measurement systems for keeping specifically the feature of flexibility. This class of works treats different measurement fields (*monitoring*), and aspects (*analysis*, *simulation*, and *design*).

About *monitoring* and *analysis*, as an example, in [43], the data acquisition hardware platform of three-dimensional machining force measurement presents a flexible software package for online monitoring, analysis, and testing. Analogously, flexible monitoring and signal processing platforms with modular software were developed for biomedical application [44, 45] and vibration analysis [46].

Regarding *simulation* and *design* aspects, a flexible radar system *simulator* applied for tank level measurement emulates the effects of antenna designs, thus to accelerate the verification process [47]. In [48], the *design* of an experimental flexible energy measurement system, consisting of distributed sensor networks with versatile and agent-based communication software is treated. A structured graphical method applied to design and implement intelligent instruments is presented in [49], where the conceptualization of the system is put in a new graphical Object-Oriented tool.

1.4.1.2 Distributed and Diagnostics Systems

In literature, a subgroup of research trends on flexible measurement systems is devoted to distributed and diagnostics-oriented measurement systems. Differently from a traditional centralized data acquisition, the approach is to distribute the devices around the specific application, in order to both interact locally in the test environment, and have a cheaper and low complex system. Research is devoted to integrated software platforms for test and diagnosis based on data services, packages, and definition of interfaces [50–52]. New software tools are applied to measure the temperature and humidity accurately with low cost [53, 54]. Particular attention is paid to the management software of the acquired data, in monitoring applications [51, 55], to automate the measurement tasks [56, 57]. The development of virtual instrumentation is presented in [58], applied to torsional vibration and phasor measurement, respectively. A platform for simulation and real-time autonomous guided vehicles navigation [59] employs software architecture and code to reduce development time necessary for debugging, optimizing control algorithms, and identifying system.

Web-based

A further significant research trend is devoted to distributed measurement systems classified as web-based (Figure 1.3), owing to their feature of remote control and monitoring by the internet. In this class, other sub-trends can be identified, according to the application (*integrated laboratory*, and *industrial process*), software programming (*Object Oriented*), and environment (*WWW-based*). In [60–63], *integrated laboratory* environments are aimed at providing remote access to heterogeneous equipment for a multiplicity of users. Web tools, based on *Object-Oriented* programming and client/server communications, have been developed for allowing remote configuration and flexible management of remote instruments [64–66]. In [67] and [68], *WWW-based* software environments are used for the design of panels of virtual measuring instruments, and for measurement data access, respectively. Integrated systems, again based on the web, are applied to remote monitoring and control of *industrial processes* [69].

1.4.2   SPECIFIC AND CUSTOM SOFTWARE

In the state of the art of research, often software applied to measurement is difficult to identify, and most of the times, it's hidden in the hardware description. From this point of view, another category of software application, *Specific and Custom Software*, is identified (Figure 1.3). In this category, the software applied to measurement systems, which cannot be considered as platforms or flexible systems, is included. However, by analyzing the literature, different technical features of these specific and custom systems can be distinguished. The technical features are referred to the following particular application system: (1) *Automatic and Control Systems*, (2) *Sensor Networks*, and (3) *Dedicated Devices*. In the following, a general discussion of these software applications is presented.

1.4.2.1 Automatic and Control Systems

Among the specific and custom software, a first significant category is represented by software applied to automatic and control systems. The nature of a control system of a process involves first a comparison between the current measurements and a reference value, and, then, in presence of differences, a suitable feedback response. The main feature of this kind of systems is the specific application to be controlled, and from this, the relative software is specific and customized to the control application. An additional sub-classification can be carried out according to the control field (*energy, transmission, motion, monitoring,* and *graphical*).

About *energy* control software, typical significant examples are virtual instrument software, developed for automatic measurement systems applied to power systems monitoring [70], or to battery resistance measurement [71]. Moreover, a multifunctional virtual instrument system for harmonic measurement of voltage and current signals is designed and implemented in [72].

In the *transmission* control field, an UHF radio-frequency identification tag test and measurement system based on virtual instrument programming is proposed in [73]. In a power flow controller of AC transmission system, a software development system for measurement and control field is employed for the system application [74].

For measurement and control on high-precision *motion* platforms [75, 76], software systems are implemented to controlling the main components of high-precision system. Based on the principle of virtual instrument, a measurement and control system for sling stretch test machine was constructed via the software platform [77], and for motion control of pipeline inspection robot [78].

For condition *monitoring* application, the design and development of the data-acquisition and storage parts of a measurement system include pc-based software implementation of virtual instruments [79, 80].

Finally, a *graphical* software tool with user interface was implemented to assist the selection of extrapolation methods for moving-boat ADCP streamflow measurements [81].

1.4.2.2 Sensor Networks

A sensor network consists of spatially distributed autonomous sensors for monitoring physical and environmental conditions. Most modern networks are wireless and bidirectional, also to enable complex plants control, and are used in many industrial and consumer applications, such as health monitoring, industrial process, machine control, and so on. Similar to the previous class of automatic and control systems, the sensor networks depend strongly on the field application, and on the particular case of study. For this reason, the related software can be considered as specific and case study-oriented.

For continuous monitoring with multisensor data acquisition GPS-based technology, the measurement techniques are based on general-purpose [82] and on embedded software [83]. In [84], a temperature measurement and control system for constant temperature reciprocator platelet preservation box is designed based on Fuzzy-PID control, and virtual instrument software. A Multipoint Wireless Measurement System is presented with multiple sensors transmission and virtual instrument interface for data processing [85]. A virtual instrumentation support system that permits to run several concurrent virtual instruments has been developed like a multitasking graphical environment [86].

1.4.2.3 Dedicated Devices

The class of software for dedicated devices includes all the packages designed and conceived for custom devices set apart for special applications. In this class, different purposes or trends can be identified, such as *integration, control, flexible* or *user friendly* features, *general measurement,* and *monitoring systems.*

About the *integration* and *control,* for the online diagnosis of reactive plasmas, a flexible data acquisition and automatic control system based on virtual instrument programming [87] was designed to control and integrate all the stand-alone measurement instruments including a spectrometer, a high-performance oscilloscope, a laser system, and a digital delay generator into a single personal computer-based control unit. In [88], for the Ionospheric Bubble Seeker, a new post processing technique based on the Java programming language was developed for all the operating systems and allowing also a remote control. Also *integration*-oriented are the virtual instruments presented in [89, 90], the former is an auto-measurement system of wave-plates phase retardation, and the former a software for tests of the Large Binocular Telescope. In [91], the *control* software of an ultrasonic sensor for pressure measurements is constituted by a virtual instrument.

About the *flexibility* aspect and the capability of making *user friendly* a particular hardware, typical examples are related to a virtual instrument integrated with a power network analyzer [92] or the implementation of the measurement communication of a weather station [93].

Other dedicated devices are inserted in traditional *measurement* and *monitoring systems*, such as the system applied to the leakage current of insulators using a virtual instrument for analyzing the data [94], and the grounding measurement system of substations adopting a virtual instrument to implement the small electric current method [95].

Finally, in [96], custom software is implemented for measuring RF chip properties and for simulating the system.

1.4.3 APPLICATION FIELD

The objective of this section is to highlight the main application fields touched by the development of new measurement software in research literature. Starting from the aforementioned analysis (*Platform and Specific Software*) of the state of the art, *Electronics* [39, 46, 47, 57, 61–63, 72, 73, 75, 86, 92, 96], represent the main investigation topic, and thus, the first and main subject in the tree (Figure 1.3) is logically the development of electronics applied to measurement. Main contributions can be found in *Diagnostics* [25, 33, 36, 37, 40, 50, 52–54, 69, 78, 83], *Physics* [20, 22, 27, 28, 30, 32, 41, 68, 75, 87] and *Energy* [38, 48, 58, 71, 79, 80, 82, 85, 94, 95] fields.

During the last decade, the main innovation demands have occurred in the field of automatic control and industrial measurement for enhancement of the product quality. Regarding world research centers, the growth in number and size of the research centers has prompted the development of new software for the management of the procedures and methods of measurement. In the field of *Energy*, the greatest stimulus for the development of new software has been given by the growing economic interests and the need for managing the distribution network. Other promising fields where the measurement–oriented software has grown are represented by *Biology* [21, 56, 84], *Medicine* [24, 35, 44, 45], and *Mechanics* [59, 76, 77, 91].

1.4.4 SOFTWARE ENVIRONMENTS

In literature, different platforms or programming languages, used for developing the measurement applications, can be identified. In first analysis, main consideration regards the impossibility of finding a direct link between technical software features and the used software platform. However, two main trends can be identified: the former is related to the transversal use of user-friendly platforms, and the latter to the link between the nature of the measurement application and the choice of the implementation environment. For the first trend, the graphical programming language (LabVIEW of NI, in Figure 1.3) shows usability greater than the Object-Oriented programming in all the custom applications and in few flexible measurement systems. This aspect depends both on the typology of measurement application, and on the possibility of reusing the ready-implemented software. From this point of view, the commercial software platforms, such as LabVIEW, are advantaged with respect to the custom platforms to provide programming developed from scratch. About this second trend, the choice of implementation platform depends strictly on the size and type of measurement application. In the research project application of large size and particular typology, the trend of using the Object-Oriented Programming can be identified.

REFERENCES

[1] Bosch, J. 1998. "Design of an Object-Oriented Framework for Measurement Systems." *Object-Oriented Application Frameworks Conference.*

[2] National Instruments. 2013. *Choosing the Right Software Application Development Environment (ADE)*, http://www.ni.com/white-paper/3091/en/

[3] National Instruments. 2014. *Homepage*, http://www.ni.com/

[4] National Instruments. 2014. *LabVIEW*, http://www.ni.com/LabVIEW/

[5] National Instruments. 2014. *Measurement Studio*, http://www.ni.com/mstudio/

[6] Agilent Technologies. 2014. *Homepage*, http://www.agilent.com/home

[7] Agilent Technologies. 2014. *Liquid Chromatography*—Instrument Control Framework, http://www.chem.agilent.com/en-US/products-services/Instruments-Systems/Liquid-Chromatography/Pages/lcf.aspx

[8] Agilent Technologies. 2005. *Agilent OpenLAB Laboratory Software Framework*, http://www.chem.agilent.com/Library/datasheets/Public/5989-3712en_lores.pdf

[9] Agilent Technologies. 2013. *Agilent VEE Pro*, http://cp.literature.agilent.com/litweb/pdf/5990-9117EN.pdf

[10] Agilent Technologies. 2014. *BenchVue Software*, http://www.home.agilent.com/en/pd-2368912-pn-34840B/benchvue-software

[11] Azimuth Systems. 2014. *Homepage*, http://www.azimuthsystems.com/

[12] Azimuth Systems. 2014. *Director II Test Management Software*, http://www.azimuthsystems.com/products/azimuth-director/

[13] EADS North America Test and Services. 2014. *Homepage*, http://www.ts.eads-na.com/

[14] EADS North America Test and Services. 2014. *Software*, http://www.ts.eads-na.com/software

[15] EADS North America Test and Services. 2013. *ATLAS*, http://www.eads-tes.co.uk/Standards/ATLAS/ATLAS.htm

[16] EADS North America Test and Services. 2014. *SigBase*, http://www.ts.eads-na.com/software/sigbase

[17] U.S. Department of State—Directorate of Defense Trade Controls. 2014. *The International Traffic in Arms Regulations (ITAR)*, http://www.pmddtc.state.gov/regulations_laws/itar.html

[18] Stenvard, P., A. Hansebacke, and N. Keskitalo. May 1–3, 2007. "Considerations When Designing and Using Virtual Instruments as Building Blocks in Flexible Measurement System Solutions." *Instrumentation and Measurement Technology Conference Proceedings, IMTC 2007*, Vol. 6 (Session 2), pp. 1–5. Warsaw, Poland: IEEE.

[19] Bitoleanu, A., M. Popescu, E.G. Subtirelu, and O.D. Dobriceanu. September 2009. "Electrical Stations Real-time Monitoring Using Combined Protection and Control Systems." *IEEE International Symposium on Diagnostics for Electric Machines, Power Electronics and Drives, (SDEMPED 2009)*, pp. 1–6. Cargese, France: IEEE.

[20] Arpaia, P., M. Buzio, L. Fiscarelli, and V. Inglese. July 2012. "A Software Framework for Developing Measurement Applications under Variable Requirements." *Review of Scientific Instruments* 83, no. 11, p. 115103. doi: http://dx.doi.org/10.1063/1.4764664

[21] Langer, D., M. Van't Hoff, C. Nagaraja, A.J. Keller, O.A. Pfaffli, M. Goldi, H. Kasper, and F. Helmchen. February 2013. "HelioScan: A Software Framework for Controlling In Vivo Microscopy Setups with High Hardware Flexibility, Functional Diversity and Extendibility." *Journal of Neuroscience Methods* 215, no. 5, pp. 38–52. doi: http://dx.doi.org/10.1016/j.jneumeth.2013.02.006

[22] Gateau, M., M. Marchesotti, A. Raimondo, A. Rijllart, and H. Reymond. September 2005. "Experience with Configurable Acquisition Software for Magnetic Measurement Experience." *14th International Magnetic Measurement Workshop (IMMW14)*, Geneva, Switzerland: CERN.

[23] Leung, Y., J.H. Ma, and M.F. Goodchild. December 2004. "A General Framework for Error Analysis in Measurement-Based GIS Part 4: Error Analysis in Length and Area Measurements." *Journal of Geographical Systems* 6, no. 4, pp. 403–28. doi: 10.1007/s10109-004-0141-4

[24] Armitage, P.A., C.S. Rivers, B. Karaszewski, R.G.R. Thomas, G.K. Lymer, Z. Morris, and J.M. Wardlaw. March 2012. "A Grid Overlay Framework for Analysis of Medical Images and Its Application to the Measurement of Stroke Lesions." *European Radiology* 22, no. 3, pp. 625–32. doi: 10.1007/s00330-011-2284-2

[25] Giachetti, R.E., L.D. Martinez, O.A. Saenz, and C.S. Chen. October 2003. "Analysis of the Structural Measures of Flexibility and Agility Using a

Measurement Theoretical Framework." *International Journal of Production Economics* 86, no. 1, pp. 47–62. doi: 10.1016/S0925-5273(03)00004-5

[26] Bondavalli, A., A. Ceccarelli, L. Falai, and M. Vadursi. April 2010. "A New Approach and a Related Tool for Dependability Measurements on Distributed Systems." *IEEE Transactions on Instrumentation and Measurement* 59, no. 4, pp. 820–31. doi: 10.1109/TIM.2009.2023815

[27] Shen,, X., X. Song, and J. Chen. August 2009. "Implementation and Evaluation of Object-Oriented Flexible Measurement System." *Electronic Measurement & Instruments, ICEMI'09, 9th International Conference,* pp. 3–310, 3–314. Beijing, China: IEEE.

[28] Deniau, L. September 2005. "Experience with Field Quality Analysis Software and Future Projects." *14th International Magnetic Measurement Workshop (IMMW14)*, Geneva, Switzerland: CERN.

[29] Yang, Q., and C. Butler. February 1998. "An Object-Oriented Model of Measurement Systems." *IEEE Transactions on Instrumentation and Measurement* 47, no. 1, pp. 104–7. doi: 10.1109/19.728800

[30] Nogiec, J.M., J. DiMarco, S. Kotelnikov, K. Trombly-Freytag, D. Walbridge, and M. Tartaglia. June 2006. "A Configurable Component-Based Software System for Magnetic Field Measurements." *IEEE Transactions on Applied Superconductivity* 16, no. 2, pp. 1382–5. doi: 10.1109/TASC.2005.869672

[31] Xiao-liang, X., L.Y. Wang, and H. Zhou. September 22–25, 2003. "An Object-Oriented Framework for Automatic Test Systems." *AUTOTESTCON 2003, IEEE Systems Readiness Technology Conference Proceedings*, ISSN: 1080-7725, pp. 407–10, Anaheim, CA: IEEE.

[32] Arpaia, P., L. Fiscarelli, G. La Commara, and C. Petrone. December 2011. "A Model-Driven Domain-Specific Scripting Language for Measurement-System Frameworks." *IEEE Transactions on Instrumentation and Measurement* 60, no. 12, pp. 3756–66. doi: 10.1109/TIM.2011.2149310

[33] Nogiec, J.M. 2007. "Architecture and Features of an Extensible Measurement System Framework." *15th International Magnetic Measurement Workshop (IMMW15)*, Batavia, Illinois, USA: Fermi National Accelerator Lab (FNAL).

[34] Rasche, A., and A. Polze. May 14–15, 2003. "Configuration and Dynamic Reconfiguration of Component-based Applications with Microsoft .NET." *Object-Oriented Real-Time Distributed Computing 2003, Sixth IEEE International Symposium*, pp. 164–71. Hakodate, Japan: IEEE.

[35] Gupta, R., and J.N. Bera. December 1–2, 2012. "A Framework for Cardiac Patient Monitoring Using an Intelligent Wireless System for Rural Healthcare in India." *First International Conference on Intelligent Infrastructure the 47th Annual National Convention at Computer Society of India (CSI - 2012)*, Kolkata, India: Mc Graw Hill.

[36] Chen, P.H. September 21–23, 2011. "Smart Browser: Network Measurement System Based on perfSONAR Framework." *Network Operations*

and Management Symposium (APNOMS), 2011 13th Asia-Pacific, pp. 1–4. Taipei, Taiwan: IEEE.

[37] Brusey, J., E.I. Gaura, D. Goldsmith, and J. Shuttleworth. November 2009. "FieldMAP: A Spatiotemporal Field Monitoring Application Prototyping Framework." *IEEE Sensors Journal* 9, no. 11, pp. 1378–90. doi: 10.1109/JSEN.2009.2021799

[38] Atitallah, R.B., E. Senn, D. Chillet, M. Lanoe, and D. Blouin. February 2013. "An Efficient Framework for Power-Aware Design of Heterogeneous MPSoC." *IEEE Transactions on Industrial Informatics* 9, no. 1, pp. 487–501. doi: 10.1109/TII.2012.2198657

[39] Deshmukh, A., F. Ponci, A. Monti, L. Cristaldi, R. Ottoboni, M. Riva, and M. Lazzaroni. April 24–27, 2006. "Multi Agent Systems: An Example of Dynamic Reconfiguration." *IMTC 2006 – Instrumentation and Measurement Technology Conference Sorrento*. Italy: IEEE.

[40] Horak, G., D. Vasic, and V. Bilas. May 1–3, 2007. "A Framework for Low Data Rate, Highly Distributed Measurement Systems." *Instrumentation and Measurement Technology Conference - IMTC 2007*, pp. 1–4. Warsaw, Poland: IEEE.

[41] Nogiec, J.M., and K. Trombly-Freytag. September 27–October 1, 2004. "A Dynamically Reconfigurable Data Stream Processing System." *Computing in High-Energy Physics (CHEP '04)*, pp. 429–32. Interlaken, Switzerland.

[42] Xiao-liang, X., W. Le-yu, and Z. Hong. September 22–25, 2003. "Framework Design and Implementation for Virtual Instrument Component Library of GPP." *Proceedings IEEE Systems Readiness Technology Conference AUTOTESTCON 2003*, pp. 22–25. Anaheim, CA, USA: IEEE.

[43] Dong, D., X. Luo, and H. Xu. July 2011. "Development of Flexible Three-dimensional Machining Force Measurement and Analysis System." *Mechanic Automation and Control Engineering (MACE)*. Hohhot, North China: IEEE.

[44] Camacho, J., B. Yelicich, L. Moraes, A. Biestro, and C. Puppo. September 2010. "Development of a Multimodal Monitoring Platform for Medical Research." *Conf Proc IEEE Engineering in Medicine and Biology Society 2010, 32nd Annual International Conference*, pp. 2358–61. Buenos Aires, Argentina: IEEE.

[45] Edström, U., J. Skönevik, T. Bäcklund, and J.S. Karlsson. September 1–4, 2005. "A Flexible Measurement System for Physiological Signals in Mobile Health Care." *Proceedings of the 2005 IEEE Engineering in Medicine and Biology 27th Annual Conference*, pp. 2161–62. Shanghai, China: IEEE.

[46] Vişan, D.A. and I.B. Cioc. May 2010. "Virtual Instrumentation Application for Vibration Analysis in Electrical Equipments Testing." *2010 33rd International Spring Seminar on Electronics Technology (ISSE)*, pp. 216–9. Warsaw, Poland: IEEE.

[47] Zietz, C., E. Denicke, and I. Rolfes. October 2009. "A Flexible System Simulator for Antenna Performance Evaluation of Radar Level Measurements." *EuRAD 2009*, pp. 513–6. Rome, Italy: IEEE.

[48] Driesen, J., G. Deconinck, J. Van Den Keybus, B. Bolsens, K. De Brabandere, K. Vanthournout, and R. Belmans. May 21–23, 2002. "Development of a Measurement System for Power Quantities in Electrical Energy Distribution Systems." *IEEE Instrumentation and Measurement Technology Conference Anchorage*, Vol. 19. Anchorage, Alaska, USA: IEEE.

[49] Dapoigny, R., P. Barlatier, E. Benoit, and L. Foulloy. July 29–31, 2003. "An Ontology-Based Graphical Tool for Intelligent Instruments." *CIMSA 2003 - International Symposium on Computational Intelligence for Measurement Systems and Applications*, pp. 150–5. Lugano, Switzerland: IEEE.

[50] Xu, B., L.W. Guo, and J.S. Yu. December 2012. "Software Platform for General Purpose Test and Diagnosis." *Applied Mechanics and* 1241-4, pp. 284–7. doi: 10.4028/www.scientific.net/AMM.241-244.284.

[51] Abadi, D.J., D. Carney, U. Çetintemel, M. Cherniack, C. Convey, S. Lee, M. Stonebraker, N. Tatbul, and S. Zdonik. 2003. "Aurora: A New Model and Architecture for Data Stream Management." *The VLDB Journal* 12, no. 2, pp. 120–39. doi: 10.1007/s00778-003-0095-z

[52] Angelov, P., V. Giglio, C. Guardiola, E. Lughofer, and J.M. Luján. 2006. "An Approach to Model-Based Fault Detection in Industrial Measurement Systems with Application to Engine Test Benches." *Measurement Science and Technology* 17, no. 7, pp. 1809–18. doi: 10.1088/0957-0233/17/7/020

[53] Ma, Y., J. Zhou, D. Pan, Y. Peng, and X. Peng. December 2012. "A Novel and Intelligent Integrated-Distributed Measurement Platform for Multisensors." *Instrumentation, Measurement, Computer, Communication and Control (IMCCC), 2012 Second International Conference*, pp. 266–71. Hangzhou, China: IEEE.

[54] Manuel, A., J. Del Rio, S. Shariat, J. Piera, and R. Palomera. May 2005. "Software Tools for a Distributed Temperature Measurement Systems." *IMTC 2005*, Vol. 2, pp. 1566–70. Ottawa, Canada: IEEE.

[55] Sorribas, J., J. del Río, E. Trullols, and A. Mànuel-Làzaro. October 2006. "A Meteorological Data Distribution System Using Remote Method Invocation Technology." *IEEE Transactions on Instrumentation and Measurement* 55, no. 5, pp. 1794–803. doi: 10.1109/TIM.2006.881572

[56] Drakakis, E.M., Y. Hua, M. Lim, A. Mantalaris, N. Panoskaltsis, A. Radomska, C. Toumazou, and T. Cass. May 2007. "An On-line, Multi-Parametric, Multi-Channel Physicochemical Monitoring Platform for Stem Cell Culture Bioprocessing." *ISCAS 2007*, pp. 1215–8. New Orleans, LA, USA: IEEE.

[57] Wenyue, X., and Y. Haiwen. August 16–19, 2011. "A Development Platform for Complex Data Acquisition System." *ICEMI'2011, The Tenth*

International Conference on Electronic Measurement & Instruments, Vol. 1, pp. 321–4. Chengdu, China: IEEE.

[58] Laverty, D.M., R. Best, P. Brogan, I. Al-Khatib, L. Vanfretti, and D.J. Marrow. April 2013. "The OpenPMU Platform for Open-Source Phasor Measurements." *IEEE Transactions on Instrumentation and Measurement* 62, no. 4, pp. 701–9. doi: 10.1109/TIM.2013.2240920

[59] Baglivo, L., M. De Cecco, F. Angrilli, F. Tecchio, and A. Pivato. June 1–4, 2005. "An Integrated Hardware/Software Platform for Both Simulation and Real-Time Autonomus Guided Vehicles Navigation." *Proceedings 19th European Conference on Modelling and Simulation, ECMS 2005*, Riga, Latvia: ResearchGate.

[60] Billaud, M., D. Geoffroy, P. Cazenave, and T. Zimmer. October–December, 2009. "A Distance Measurement Platform Dedicated to Electrical Engineering." *IEEE Transactions on Learning Technologies* 2, no. 4, pp. 312–9. doi: http://dx.doi.org/10.1109/tlt.2009.45

[61] Davoli, F., G. Spanò, S. Vignola, and S. Zappatore. October 2006. "LAB-NET: Towards Remote Laboratories with Unified Access." *IEEE Transactions on Instrumentation and Measurement* 55, no. 5, pp. 1551–8. doi: http://dx.doi.org/10.1109/tim.2006.880919

[62] Arpaia, P., F. Cennamo, P. Daponte, and M. Savastano. June 4–6, 1996. "A Distributed Laboratory Based on Object-Oriented Measurement Systems." *IEEE Instrumentation and Measurement Technology Conference*, Vol. 19 (Session 3), pp. 207–15. Brussels, Belgium: IEEE.

[63] Tawfik, M., E. Sancristobal, M. Sergio, R. Gil, G. Diaz, A. Colmenar, K. Nilsson, J. Zackrisson, L. Håkansson, and I. Gustafsson. January–March, 2013. "Virtual Instrument Systems in Reality (VISIR) for Remote Wiring and Measurement of Electronic Circuits on Breadboard." *IEEE Transactions on Learning Technologies* 6, no. 1, pp. 60–72. doi: http://dx.doi.org/10.1109/tlt.2012.20

[64] Grimaldi, D., L. Nigro, and F. Pupo. February, 1998. "Java-Based Distributed Measurement Systems." *IEEE Transactions on Instrumentation and Measurement*, Vol. 47, no. 1, pp. 100-3. doi: 10.1109/19.728799.

[65] Bertocco, M., S. Cappellazzo, A. Carullo, M. Parvis, and A. Vallan. June, 2003. "Virtual Environment for Fast Development of Distributed Measurement Applications." *IEEE Transactions on Instrumentation and Measurement* 52, no. 3, pp. 681–5. doi: http://dx.doi.org/10.1109/vims.2001.924901

[66] Pianegiani, F., D. Macii, and P. Carbone. June 2003. "An Open Distributed Measurement System Based on an Abstract Client-Server Architecture." *IEEE Transactions on Instrumentation and Measurement* 52, no. 3, pp. 686–92. doi: http://dx.doi.org/10.1109/tim.2003.814699

[67] Michal, K., and W. Wieslaw. May 21–23, 2001. "A New Java-Based Software Environment for Distributed Measurement Systems Designing." *IEEE Instrumentation and Measurement Technology Conference* 51, no. 6, pp. 1340–6. doi: http://dx.doi.org/10.1109/imtc.2001.928847

[68] Nideröst, B., M. van de Giessen, W. Lourens, and J. Krom. June 2002. "The WebUmbrella Web-Based Access to Distributed Plasma-Physics Measurement Data." *IEEE Transactions on Nuclear Science* 49, no. 3, pp. 1579–83. doi: http://dx.doi.org/10.1109/tns.2002.1039703

[69] Nikolakopoulos, G., M. Koundourakis, and A. Tzes. July 21–29, 2003. "An Integrated System Based on WEB and/or WAP Framework for Remote Monitoring and Control of Industrial Processes." *VECIMS 2003 - International Symposium on Virtual Environments, Human-Computer Interfaces, and Measurement Systems*, pp. 201–6. Lugano, Switzerland: IEEE.

[70] Bai, Y.W., and W.-C. Kuo. November, 2010. "Design and Implementation of an Automatic Measurement System for DC-DC Converter Efficiency on a Motherboard." *IECON 2010*, pp. 1323–8. Phoenix, AZ, USA: IEEE.

[71] Yongjie, F. August, 2011. "Design of the Battery Resistance Measurement System." *Electronic Measurement & Instruments (ICEMI) 2011*, Vol. 2, pp. 240–3. Chengdu, China: IEEE.

[72] Tang, Q., Y. Wang, and S. Guo. October 18–20, 2008. "Design of Power System Harmonic Measurement System Based on LabVIEW." *Natural Computation, 2008. ICNC '08. Fourth International Conference*, Vol. 5, pp. 489–93, Washington, DC, USA: IEEE.

[73] Nikitin, P.V., and K.V. Seshagiri Rao. July 2009. "LabVIEW-based UHF RFID Tag Test and Measurement System." *IEEE Transactions on Industrial Electronics* 56, no. 7, pp. 2374–81. doi: http://dx.doi.org/10.1109/tie.2009.2018434

[74] Chen Y. March, 2009. "Electric Quantity Test System of Unified Power Flow Controller Model on LabWindows/CVI." *APPEEC 2009*, Wuhan, China: IEEE.

[75] Batusov, V.Y., M.V. Lyablin, and N.D. Topilin. May 2011. "Development and Application of the Complex Hardware/Software System for Controlled Assembly of the ATLAS Hadron Tile Calorimeter." *Physics of Particles and Nuclei* 42, no. 3, pp. 438–59. doi: http://dx.doi.org/10.1134/s1063779611030026

[76] Xiaolong, K., G. Yinbiao, and G. Longxing. April 2009. "Study on Software and Hardware Control of High-precision Measurement Platform for Optical Aspheric Surface." *2009 International Conference on Measuring Technology and Mechatronics Automation*, Zhangjiajie, China: IEEE.

[77] Chen, C., and F. Hu. May 30–31, 2010. "Design of Measurement and Control System for Sling Stretch Test Machine Based on LabVIEW." *2010 2nd International Conference on Industrial Mechatronics and Automation, ICIMA2010*, Vol. 2, pp. 389–92. Wuhan, China: IEEE.

[78] Cao, J., L. Lin, Y. Li, T. Huo, and F. Dai. August 8–10, 2011. "The Measurement and Control System for Pipe Inspection Robot Based on LabVIEW." *Artificial Intelligence, Management Science and Electronic Commerce (AIMSEC), 2011 2nd International Conference*, pp. 4497–500. Deng Feng, China: IEEE.

[79] Wang, P., X. Liu, and Y. Han. December 2012. "A Single-phase Electric Harmonic Monitoring System Design Based on the LabVIEW." *Advanced Materials Research*, Vol. 605–7, pp. 664–8. doi: 10.4028/www.scientific.net/AMR.605-607.664.

[80] Poza, F., P. Mariño, S. Otero, and F. Machado. April 2006. "Programmable Electronic Instrument for Condition Monitoring of In-Service Power Transformers." *IEEE Transactions on Instrumentation and Measurement* 55, no. 2, pp. 625–34. doi: http://dx.doi.org/10.1109/tim.2006.870122

[81] Mueller, D.S. January 2013. "extrap: Software to Assist the Selection of Extrapolation Methods for Moving-boat ADCP Streamflow Measurements." *Computers & Geosciences* 54, pp. 211–8. doi: 10.1016/j.cageo.2013.02.001

[82] Locci, N., C. Muscas, and S. Sulis. November 2008. "A Flexible GPS-Based System for Synchronized Phasor Measurement in Electric Distribution Networks." *IEEE Transactions on Instrumentation and Measurement* 57, no. 11, pp. 2450–6. doi: http://dx.doi.org/10.1109/tim.2008.924930

[83] Li, S., J. Tian, Z. Yang, and F. Qiao. March 2013. "Research and Implement of Remote Vehicle Monitoring and Early-Warning System Based on GPS/GPRS." *International Conference on Graphic and Image Processing (ICGIP 2012)*, Singapore: SPIE.

[84] Jiang, X., L. Zhang, and H. Xue. 2008. "Designing a Temperature Measurement and Control System for Constant Temperature Reciprocator Platelet Preservation Box Based on LabVIEW." *Fourth International Conference on Natural Computation*, Jinan, China: IEEE.

[85] Yang Z., X. Hu, and G. Chang. May 25–27, 2012. "Research of Torsional Vibration Monitoring Platform for Turbine Generator." *Computer Science and Automation Engineering (CSAE), 2012 IEEE International Conference*, pp. 577–80. Zhangjiajie, China: IEEE.

[86] Brito Palma, L.F.F., and A.R.F. da Silva. 1998. "A Virtual Instrumentation Support System." *Electronics, Circuits and Systems, 1998 IEEE International Conference on, Lisbon, Portugal*, Vol. 1, pp. 301–4. Lisboa, Portugal: IEEE.

[87] Feng, C.L. June, 2011. "Laser Mass-spectrometry for Online Diagnosis of Reactive Plasmas with Many Species." *Review of Scientific Instruments* 82, no. 6, p. 063110.

[88] Magdaleno, S., M. Herraiz, and S.M. Radicella. May 2012. "Ionospheric Bubble Seeker: A Java Application to Detect and Characterize Ionospheric Plasma Depletion from GPS Data." *IEEE Transactions on Geoscience and Remote Sensing* 50, no. 5, pp. 1719–27. doi: http://dx.doi.org/10.1109/tgrs.2011.2168965

[89] Yanhao, L., Q. Xiaosheng, W. Dongguang, L. Jiaben, and H. Keliang. August 24–26, 2010. "Research on Auto Measurement System of Phase Retardation of Wave Plates Based on LabVIEW." *2010 International Conference on Computer, Mechatronics, Control and Electronic Engineering (CMCE)*, Vol. 3, pp. 29–32. Changchun, China: IEEE.

[90] Polsterer, K.L., M. Jütte, V. Knierim, M. Lehmitz, and H. Mandel. June 27, 2006 "Lucifer VR: A Virtual Instrument for the LBT." *Advanced Software and Control for Astronomy*, Vol. 6274. Orlando, Florida, USA: SPIE.

[91] Wenhai, H. August 2007. "Design of the Measurement System of the Pump Based on LabVIEW." *The Eighth International Conference on Electronic Measurement and Instruments ICEMI'2007* 2, pp. 375–78, 16–18. Xian, China: IEEE.

[92] Kaminsky, D. September 2010. "Modular and Flexible Power Network Analyzer." *Conference on Applied Electronics (AE) 2010*, pp. 1–4. Pilsen, Czech Republic: IEEE.

[93] Branzila, M., F. Mariut, and D. Petrisor. October 25–27, 2012. "Virtual Instrument Developed for Adcon Weather Station." *2012 International Conference and Exposition on Electrical and Power Engineering (EPE 2012)*, pp. 853–6. Iasi, Romania: IEEE.

[94] Chunhua, F., W. Jianguo, L. Yang, C. Junjie, X. Nianweng, S. Zhen, and Z. Mi. November 9–12, 2008. "Composite Insulators Leakage Current Measurement System Based on LabVIEW." *2008 International Conference on High Voltage Engineering and Application*, Chongqing, China: IEEE.

[95] Liu, Z., J. Gao, and S. Zhang. October 10–11, 2009. "Research of the Grounding Measurement System Based on LabVIEW in Substation." *2009 Second International Conference on Intelligent Computation Technology and Automation*, Vol. 2, pp. 328–31. Changsha, Hunan, China: IEEE.

[96] Jung, S., Y. Eo, Y. Chun, W. Kim, S. Woo, S. Sohn, J. Rho, and J. Cha. September 28–30, 2011. "UWB Sensor Chip Measurement System Implementation Using Labview and MCU Board." *2011 International Conference on ICT Convergence (ICTC)*, pp. 649–51. Seoul, Korea: IEEE.

CHAPTER 2

SOFTWARE FRAMEWORKS FOR MEASUREMENT APPLICATIONS

If you want to teach people a new way of thinking, don't bother trying to teach them.
Instead, give them a tool, the use of which will lead to new ways of thinking.

—Richard Buckminster Fuller

2.1 OVERVIEW

In this chapter, the general topic of software frameworks is presented and the main related concepts are introduced. Subsequently, these principles are examined in the more specific case of software frameworks for measurement applications, by discussing their rationale and main features. Then, a further section is dedicated to introduce the related theme of Domain Specific Languages that are vital for software frameworks. Finally, the requirements and the desirable features of a software framework for measurement applications are summarized in relation to the different types of users who would interact with such a system.

2.2 GENERAL CONCEPTS

In software development, a framework is a well-defined support structure for organizing and developing a software project. The user is provided with more or less integrated (but surely helpful) tools and interfaces in order to facilitate the development of projects. A framework may include support programs, libraries, a scripting language, and other software tools to help develop and link the different components of a software project.

The design of a framework usually addresses issues not corresponding to the final applications produced through the framework itself. Frameworks are realized with the aim of facilitating software development usually abstracting to the single specific application. The main objective is to allow final users to spend more time on meeting the application requirements, rather than dealing with the low level details for the implementation of a working system. Framework design should be, as much as possible, general and independent on the details of a specific target application, so it can be easily adapted to the needs of new applications, which were not originally foreseen at an early development stage. At the same time, a suitable design approach should be employed in order to allow the resulting framework to effectively produce high-quality software for each target application.

From a structural point of view, according to the Object-Oriented Paradigm, a framework can be seen also as a partial design and implementation of an application in a given domain [1]; it is therefore defined by a set of abstract classes and the way the instances of those classes interact. The features and the architecture of the framework can be adapted and combined to create complete applications. Thus, frameworks provide Object-Oriented systems with a higher level of reuse and allow for a considerable reduction of the effort necessary for the realization of new applications.

Frameworks should not be confused with libraries of classes, because some essential features differentiate these two categories. The main and peculiar difference is that the framework, besides containing a set of classes and interfaces, includes also a suitable infrastructure specifically aimed at a substantial reuse of the code (e.g. for managing flow control, objects interaction, and so on). In the framework operation, this feature results in a significant reduction of the amount of additional code required for the development of new applications. Furthermore, once the infrastructure has been tested and validated, its reuse will produce applications that are intrinsically robust to small-term variations in requirements, that is, flexible in the short term.

Obviously, these benefits demand in return a different approach to the development of new software applications, where the advantages of a framework are achieved in the long term. Correspondingly, the development of a framework demands a greater effort in terms of design, implementation, and also user training, when compared to a single stand-alone application, but when the number of applications to be handled grows over time and new programs have to be developed, the benefits of the framework become evident.

A traditional classification of frameworks can be found in [2], where three categories are proposed:

- *Application frameworks*, which encapsulate software expertise at a professional level applicable to a wide range of programs. These frameworks include a *horizontal* set of functions, common to professional software and independent of the application, to be exploited across the user domains, for example, Graphical User Interfaces, documents, databases, and so on.
- *Domain frameworks*, which help to implement programs for a given application domain. Frameworks of this type include a *vertical* set of functions, related to the application domain; they are specifically designed for reducing the amount of work required to develop an application in the target domain.
- *Support frameworks*, which address very-specific *system-level* applications like memory management, or file systems, and are sometimes used with the aforementioned application or domain frameworks to simplify the development of programs.

Another classification proposed in [2], is based on how a framework is used, namely, depending on whether the user can derive new classes or rather instantiate and combine already existing ones:[1]

- *Architecture-driven frameworks* use inheritance for customization. Users tailor the behavior of the framework through the derivation of new classes and the overriding of member functions.
- *Data-driven frameworks* use mainly object composition for customization. Users tailor the behavior of the framework through combinations of different objects. The combined objects affect the framework performance, but it's up to the framework itself to define how the objects can be combined.

Frameworks that can be used by inheritance only are also called *white-box frameworks,* because they cannot be extended without a deep knowledge of their internal working. Frameworks that can also be used

---

[1]This categorization can be referred to in different ways, typically as architecture-driven versus data-driven, or inheritance-focused versus composition-focused.

by configuring existing components are called *black-box frameworks,* because they provide components to be used by means of interfaces and specifications only ("black-box reuse"). Black-box frameworks are generally easier to use, because their internal mechanisms are intentionally hidden, completely or partially. In this case, the main drawback is that the adoption of a black-box reuse strategy limits the possibility of framework expansion to the set of components already provided by the framework developers [3].

In most cases, a framework does not exactly match any one of the categories presented earlier, but can be rather described as a combination of some of them. In this way, the design can be more effective in fulfilling the requirements for which the system is developed.

As an example, heavy architecture-driven frameworks might result in being difficult to use because they oblige their users to write considerable amounts of additional code to produce the desired behavior. Conversely, data-driven frameworks are generally easier to use but may show limits on achieving the desired behavior.

One approach for building frameworks both easy to use and extensible is to provide an architecture-driven base with a data-driven layer. Most frameworks provide users with twofold ways on both how to use the built-in functions and modify or extend those functions. Typically, frameworks' built-in functions are used by instantiating classes and calling their member functions. Users can extend and modify a framework's function by deriving new classes and overriding member functions. In other words, frameworks usually offer both white-box and black-box mechanisms; they have (a) a white-box layer, consisting of *interfaces* and *abstract classes,* providing the architecture to be used for white-box reuse, and (b) a black-box layer, consisting of *concrete classes and components* inherited from the white-box.

2.3 WHY A FRAMEWORK FOR MEASUREMENTS?

In the management of a test and measurement laboratory, one of the main activities is the production of suitable software for an automatic bench, satisfying a given set of test requirements. This activity is particularly costly and burdensome when the test requirements are variable in time, for example, in a small metrological laboratory in charge of calibration of bench instruments, such as multimeters, impedance meters, digital counters, digital scopes, and so on. When a batch of objects to be tested arrives, if the test is burdensome to be carried out manually, the need for developing an automatic bench arises. Test engineers define measurement requirements, and the automatic bench is designed and developed by lab technicians.

If the objects batches under test have small size, time-varying nature and composition, and frequent occurrence, the activity of bench automation becomes predominating with respect to the test execution. In this case, most significant efforts are devoted to software development.

In the last years, the problem of easy-to-assemble and -configure hardware has been faced progressively and effectively solved. Both instrumentation standard interfaces (e.g., IEC 626, VXI, PXI, and so on) and automatic circuit cabling devices (e.g., ADLINK PXI-7921 [4] and NI PXI-2529 [5]) have been successfully defined and made available on the market.

For the software, a different strategy has been set up, going through an approach of merely simplifying the programming, such as the standard de facto LabVIEW of National Instruments [6]: the graphical programming language G, using graphical icons and wires symbolizing the data flow. The approach is to emphasize the objects involved in the application and the exchange of data among them, with little care for the temporal sequences of the actions to be executed. Conversely, imperative programming languages (e.g., C, Python, and so on) point out the operation's order and allow the temporal constraint in a measurement application to be managed easily.

However, for laboratory operating with rapidly- and highly-varying test requirements, simplifying programming reduces the effort only indirectly. Moreover, the quality of the produced software, in terms of flexibility, usability, and maintainability, is not fostered intrinsically [7].

With the aim of developing test and measurement software, a particular type of domain framework is needed, conceived with measurement applications as target domain, namely, with the purpose of allowing the production of high-quality measurement applications with reduced effort [7]. Further specific necessities are to be addressed when developing an automated measurement system—for example, real-time constraints or hardware-related requirements.

Standalone test programs are designed to solve specific problems with extremely limited capability to evolve. Conversely, a framework for measurement applications is suitably conceived [2] to provide support for satisfying a wide range of requirements. This framework could constitute a unified solution to drive all the existing and future park of automatic measurement systems to be developed within an organization. As explained earlier, it would prove to be useful when a number of very specific tests are expected to be developed, rapidly adapted, and performed on single prototypes or relatively small batches of units [7].

These tests require the control of various devices, such as transducers, actuators, trigger/timing cards, power supplies, and other devices not completely specified yet. Moreover, for different measurement techniques and tests, different algorithms have to be implemented.

The ideal situation would be to have a flexible software framework, providing a robust library to remotely drive all the instrumentation involved in the tests, as well as the tools to help the user in the design of new measurement procedures for different measurement techniques. This has to be done while taking care of maximizing the measurement software quality, in terms of flexibility, reusability, maintainability, and portability, and by simultaneously keeping high efficiency levels.

2.4 DOMAIN SPECIFIC LANGUAGES

In recent years, the Model-Driven Engineering (MDE), a software development methodology creating models and abstractions closer to some particular domain concepts, has been exploited fruitfully [8]. Its main aim is to increase the software productivity, by simplifying and standardizing the process of design and implementation. For software frameworks, this paradigm finds a concrete application in Domain Specific Languages (DSLs): "MDE technologies offer a promising approach to address the inability of third-generation languages to alleviate the complexity of platforms and express domain concepts effectively" [8]. As a matter of fact, in the context of model-driven software development, the domain specific language is often exploited: MDE and DSLs are complementary and they are both necessary for a successful model-driven approach.

MDE allows application-domain variability to be handled, by providing modeling dimensions for subject areas and architectural aspects. For example, in [9], a specific kind of automation system, called test bed, used in the automotive industry for developing combustion engines, is illustrated. Due to the ever-changing measurement tasks during engine development, test beds must be extremely flexible and customizable. MDE approach allows configuration data to be derived automatically.

Kent [10] defines two additional categories of dimensions needed for MDE. The former category includes various dimensions of interest, like different subject areas and system aspects. The latter category is less concerned with the technical aspects of a system and more with organizational issues, like: authorship, version (as in version and configuration control), location (in case the system development is distributed across sites), and stakeholder (e.g., business expert or programmer).

When building models in a software development process, the model is exploited at an intersection of the dimensions. The intersecting dimensions play an important role in the choice for a modeling language (mostly indicated as a DSL [11]) for that particular model. For example, the subject area, the stakeholders, and the level of abstraction influence the modeling language. However, a DSL for a modeling language doesn't mean that it should be a visual/graphical DSL. A model is just an abstract representation of reality and it can also be expressed using a textual DSL. In short: MDE methodology defines a framework of dimensions with their intersections, and, in this way, defines also the different models needed to describe a certain software application.

According to this approach, a DSL can be defined as in [11]: *a programming language, or an executable specification language, that offers, through appropriate notations and abstractions, expressive power focused on, and usually restricted to, a particular problem domain.* Therefore, the peculiarity of DSL is its focused expressive power. A DSL can be regarded as a programming or specification language dedicated to a particular domain or problem. The benefit of a DSL with respect to a general-purpose language is the capability of providing appropriate built-in abstractions and notations. In particular, DSLs use terms derived from a model created for a specific problem domain and utilized for defining components or complete solutions to be exploited in that domain [12].

In synthesis, main advantages of DSLs are (a) to express solutions in the idiom and at the level of abstraction of the problem area; thus, domain experts can understand, validate, modify, and often even develop programs directly by themselves; (b) to enhance quality, productivity, reliability, maintainability, portability, and reusability of the software; and (c) to allow a direct validation at the domain level: the language rules can include the domain regulation.

Conversely, some drawbacks arise: cost of designing, implementing, and maintaining a new language, as well as related to the development tools, burden of learning a new language, and potential loss in efficiency compared with hand-written code.

A new language involves the development of new parser, builder, and all the tools for a profitable use. This discourages the growth of new DSLs [13]. The break-even point can be taken down if the aforementioned components and tools are already available as generic elements [14] and they have to be only specialized to the domain of interest. If DSL developers also take care about designing the constructs of the language in a simple way and with meaning related to the domain concepts, the new language will be easy to learn and exploited fruitfully by the domain operators.

Recently, practical [15, 16] and theoretical [17, 18] tools for DSL development have also been proposed, by demonstrating the increasing interest in such methods. Examples of useful exploitation of DSL concepts can be found in home automation [12], network performances testing [19], and programming of network devices [20]. Generally, a program written in a DSL is called domain-specific description (DSD) and it is compiled, interpreted, or analyzed by a domain-specific processor. After the compilation, a textual (e.g., in a general-purpose programming language) or binary format output is obtained.

Furthermore, external or internal DSLs can be defined. An external DSL is a domain-specific solution, represented in a natural-like language separate from the main programming environment. It has its own custom syntax and a specific parser (to be implemented). An internal (or embedded) DSL, instead, expresses new domain constructs within the syntax of a general-purpose language, suitably modified for this purpose.

A language is a set of terms and expressions bounded by a set of syntax and semantic rules and used for communication within a domain. General Purpose Languages (GPLs) are not specialized and are suited for a wide area of applications from business processing up to scientific computing. Conversely, DSLs are explicitly tailored to a target domain: "DSLs offer substantial gains in expressiveness and ease of use compared with GPLs in their domain of application" [21]. Complex constructs and abstraction of the domain are offered within the language, thus increasing its expressiveness in comparison to GPLs. The higher abstraction level, the compactness, and consequently the better readability/writability, allow expert programmers to be productive using the DSLs. This also improves productivity, because solutions for domain problems can be expressed by smaller effort, and decreases maintenance costs.

As mentioned earlier, a DSL has potential shortcomings; in fact, the main drawback is the related development effort. A DSL developer needs, at least, experience in language design and knowledge about the target domain. The developer has to find suitable abstractions and the right tradeoff between GPL and DSL constructs, as well as the language must be implemented and maintained.

Other problems are tool availability, user training costs, and performance. While GPLs have a strong tool support, the corresponding tools for a new DSL have to be created. Proper development methods and suitable tools have to be chosen in order to avoid the DSL development costs from surpassing the estimated savings derived from its usage.

Finally, a DSL might lead to a performance loss with respect to other languages. If performance is not critical, the other benefits overcome this problem. Otherwise, the developer has to pay special attention to achieve negligible performance loss.

2.5 REQUIREMENTS OF A FRAMEWORK FOR MEASUREMENT APPLICATIONS

Regarding the software, a traditional approach to the development of a measurement application often leads to limitations in the measurement control and acquisition programs, mainly associated with the relatively long time needed for development iterations (cycles composed by the steps of specification-programming-debugging-validation). As an example, measurement programs often implement a large spectrum of preprogrammed configurations that are accessible by the user, but they require software specialists for extending the set of configurations to cover new test and analysis requirements. For this reason, more advanced design principles in the field of software engineering have to be considered by introducing the concept of framework, as previously discussed in this chapter.

The framework should facilitate the development of new measurement programs, mainly by allowing easy modifications and extensions of already existing test software simultaneously. Given a set of measurement requirements to be satisfied by the available hardware, the flexible and reconfigurable platform being developed should allow an effective automatic measurement system to be generated by low cost and development time.

The resulting system, besides reproducing key operating capabilities of traditional software, has to allow user-driven and traceable configuration of the hardware as well as of the test protocols, in order to bear a maximum capability to evolve.

A framework should be characterized by (a) flexibility, for rapid and cost-effective realization of measurement applications, including prototyping in an R&D context; (b) a modular architecture, to mix and reuse components chosen from an incremental library; and (c) high performance, to exploit the increased throughput of new transducers and acquisition systems.

Frameworks should also aim at maximizing the measurement software quality, in terms of flexibility, reusability, maintainability, and portability, by simultaneously keeping high efficiency levels. In particular, the flexibility, the modification easiness of a system or component for its use

in applications or environments other than those for which it was specifically designed [7], is definitely one of the most desirable properties of any
system to face changes in operational environment during its life. This is
particularly true for software systems, both because they are often subject to extremely rapid technological development and because some of
them are specifically conceived to be employed in environments spanning
a wide range of functional requirements, not fully predictable at the design
stage. This means that a framework for measurements should be easy to
configure for satisfying a large set of measurement applications in one or
more measurement fields.

The users involved in the framework life cycle can be classified in
different categories:

- The *developer/administrator user* has suitable knowledge about
  the framework's internal structure and can access it at any level.
- The *test engineer* has main knowledge about the framework's
  features and knows functions only through their corresponding
  interfaces; he can therefore provide a formal description of the
  measurement protocol to be translated transparently into a suitable
  executable application by the framework.
- The *end user* interacts with the resulting executable measurement
  application in order to perform the tests.

The roles of the test engineer and the application user (i.e., end user)
during the exploitation of a measurement software framework are highlighted in Figure 2.1. Initially, the test engineer expresses the measurement procedure in a formal way through a script, puts this script in the
framework, and obtains an executable measurement application. In a
second phase, the measurement application user executes the corresponding measurement application, interacts with the measurement station by
providing the required input and configuring hardware setup, and finally
starts the measurement process on the devices [22].

The aforementioned roles of the different users should not be rigidly
divided. For example, if the test engineer needs to modify or add a component, the internal structure of the framework should be accessible.

The main goal of a framework for measurement applications is to
satisfy the various needs of the different users by means of both the previously mentioned white-box and black-box mechanisms, according to the
classification provided in Table 2.1.

The main responsibility of the test engineer is to interact in the easiest and most effective way with the framework through its interface in

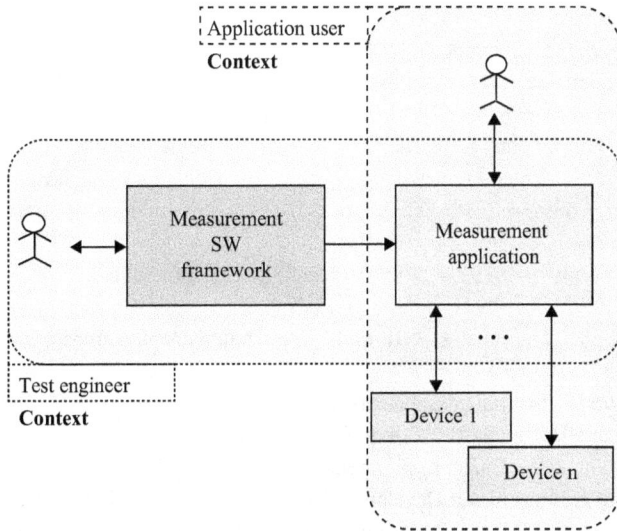

Figure 2.1. Test engineer and application user exploit a measurement software framework.
Source: Arpaia et al. [22].

Table 2.1. Main software characteristics and users they address

Software characteristic	User
Flexibility	Test engineer
Maintainability	Developer/administrator user
Reusability	Developer/administrator user, test engineer
Efficiency	Test engineer, end user

order to build the measurement applications. This can be achieved through flexibility and reusability, without necessarily including any requirements on the framework's internal organization, excluding that of producing an effective test application. This latter point is of interest for the end user performing the tests.

Conversely, the internal structure of the framework and its properties are of great interest for developers and administrator users, who will benefit from features like maintainability and reusability.

REFERENCES

[1] Bosch, J., P. Molin, M. Mattson, and P. Bengtsson. 1999. "Object-Oriented Frameworks—Problems and Expectations." In *Building Application*

Frameworks: Object-Oriented Foundation of Framework Design, eds. M.E. Fayad, D.C. Schmidt, and R.E. Johnson. London, England: Wiley and Sons.

[2] Taligent Inc. 1994. "Building Object-Oriented Frameworks." https://lhcb-comp.web.cern.ch/lhcb-comp/Components/postscript/buildingoo.pdf.

[3] van Gurp, J., and J. Bosch. 2001. "Design, Implementation and Evolution of Object-Oriented Frameworks: Concepts and Guidelines." *Software Practice and Experience* 31, no. 3, pp. 277–300. doi: http://dx.doi.org/10.1002/spe.366

[4] AD-Link. "PXI-7921." http://www.adlinktech.com/PD/marketing/Datasheet/PXI-7921/PXI-7921_Datasheet_1.pdf

[5] National Instruments. "NI PXI-2529." http://sine.ni.com/ds/app/doc/p/id/ds-375/lang/en

[6] National Instruments. January 2011. "What Is LabVIEW?" LabVIEW System Design Software. http://www.ni.com/labview/

[7] Arpaia, P., M. Buzio, L. Fiscarelli, and V. Inglese. November 2012. "A Software Framework for Developing Measurement Applications under Variable Requirements." *AIP Review of Scientific Instruments* 83, no. 11, 115103. doi: 10.1063/1.4764664.

[8] Schmidt, D.C. February 2006. "Model-Driven Engineering." *IEEE Computer* 39, no. 2, pp. 25–31. doi: http://dx.doi.org/10.1109/mc.2006.58

[9] Altendorfer, S., and H. Zsifkovits. 2013. "A Model-Driven Engineering Approach for Production Systems illustrated on an Automotive Test Case." *Proceedings in Manufacturing Systems* 8, no. 3, pp. 159–64. ISSN 2067-9238.

[10] Kent, S. May 2002. "Model Driven Engineering." *IFM '02 Proceedings of the Third International Conference on Integrated Formal Methods*, pp. 286–98. Turku, Finland: Springer Berlin Heidelberg.

[11] Fowler, M. October 2010. *Domain Specific Languages.* 1st ed. Reading, MA: Addison-Wesley.

[12] Jimenez, M., F. Rosique, P. Sanchez, B. Alvarez, and A. Iborra. July/August 2009. "Habitation: A Domain-Specific Language for Home Automation." *IEEE Software* 26, no. 4, pp. 30–38. doi: http://dx.doi.org/10.1109/ms.2009.93

[13] Hudak, P. June 1998. "Modular Domain Specific Languages and Tools." In *Software Reuse, 1998. Proceedings. Fifth International Conference*, pp. 134–42. Victoria, British Columbia, Canada: IEEE.

[14] Steinberg, D., F. Budinsky, M. Paternostro, and E. Merks. December 2008. *EMF: Eclipse Modeling Framework.* 2nd ed. Reading, MA: Addison-Wesley.

[15] Kourie, D.G., D. Fick, and B.W. Watson. February 2009. "Virtual Machine Framework for Constructing Domain-Specific Languages." *IET Software* 3, no. 1, pp. 1–13. doi: http://dx.doi.org/10.1049/iet-sen:20060068

[16] Gronback, R.C. March 2009. *Eclipse Modeling Project: A Domain-Specific Language (DSL) Toolkit.* 1st ed. Reading, MA: Addison-Wesley.

[17] Fowler, M. July/August 2009. "A Pedagogical Framework for Domain-Specific Languages." *IEEE Software* 26, no. 4, pp. 13–14. doi: http://dx.doi.org/10.1109/ms.2009.85

[18] Dib, A.A., L. Féraud, I. Ober, and C. Percebois. April 2008. "Towards a Rigorous Framework for Dealing with Domain Specific Language Families." In *Information and Communication Technologies: From Theory to Applications, 2008. ICTTA 2008. 3rd International Conference*, pp. 1–6. Damascus, Syria: IEEE.

[19] Pakin, S. October 2007. "The Design and Implementation of a Domain-Specific Language for Network Performance Testing." *Parallel and Distributed Systems, IEEE Transactions*, 18, no. 10, pp. 1436–49. doi: http://dx.doi.org/10.1109/tpds.2007.1065

[20] Muller, G., J.L. Lawall, S. Thibault, and R.E. Voel Jensen. August 2003. "A Domain Specific Language Approach to Programmable Networks." *IEEE Trans. Syst., Man, Cybern. C, Appl. Rev.* 33, no. 3, pp. 370–81. doi: http://dx.doi.org/10.1109/tsmcc.2003.817364

[21] Mernik, M., J. Heering, and A.M. Sloane. December 2005. "When and How to Develop Domain-Specific Languages." *ACM Computing Surveys (CSUR)* 37, no. 4, pp. 316–44. doi: http://dx.doi.org/10.1145/1118890.1118892

[22] Arpaia, P., L. Fiscarelli, G. La Commara, and C. Petrone. 2011. "A Model-Driven Domain-Specific Scripting Language for Measurement System Frameworks." *IEEE Transactions on Instrumentation and Measurement* 60, no. 12, pp. 3756–66. doi: http://dx.doi.org/10.1109/tim.2011.2149310

CHAPTER 3

OBJECT- AND ASPECT-ORIENTED PROGRAMMING FOR MEASUREMENT APPLICATIONS

Science and technology multiply around us.
To an increasing extent, they dictate the languages in which we speak
  and think.
Either we use those languages, or we remain mute.

—J.G. Ballard

3.1 OVERVIEW

This chapter continues our travel throughout the measurement applications software, by focusing the attention on one of the most important paradigms of the last decades: Object-Oriented Programming. The main concepts of Object-Oriented Programming and the related design patterns are presented. Finally, the main concepts of Aspect-Oriented paradigm are illustrated by emphasizing the advantage of its use in measurement applications.

3.2 OBJECT-ORIENTED PROGRAMMING

3.2.1 LEADING CONCEPTS

In the last decades, from a technological point of view, hardware and software have become increasingly complex, and their management, particularly software, turns out to be more and more a priority for the automation project engineer. Modular programming, having prevailed for several years (structured programming), focused prominently on the actions of a module (subroutine), rather than on its data. Furthermore, the subroutine was conceived as an independent logic unit aimed at reusing the code and enabling collaboration with other subroutines by means of suitable links. Researchers studied the way, by means of the so-called Object-Oriented Programming, to foster easy software production by strongly emphasizing discrete, reusable units of programming logic [1]. The main leading strategy has been to hide as much possible details about code working mechanism and low-level data structure to enhance reusability. The sense is that a logic unit (e.g., a modular procedure) is to be "plugged into" a higher-level structure easily. Details about internal working mechanism are not essential to the structure operation as a whole, provided the unit accomplishes its job properly. For example, a modular unit could include actions to configure a multimeter to trigger a measurement, wait for the measurement to end, read the results, and save the data in a file. Details about how the unit stores the measured records are hidden to the applications and systems using this module. Analogously, the programming details of the various procedures are also obscured (principle of "information hiding" of software engineering, while the mechanism for restricting access to some of the object's components is called "encapsulation"). For the sake of simplicity, only the behavior (interface) of the existing procedures and the corresponding accessible data is made available. Object-Oriented Programming focuses on data rather than processes, with procedures constituted of independent generic modules ("classes"), whose instance ("objects") contains all the information needed to manipulate its own data structure ("members") [2].

In an Object-Oriented program, different types of objects correspond to different kinds of complex data to be managed, and each of them corresponds to a real-world object or concept (device, person, etc.). Multiple copies of each type of object might be contained in a generic program.

Concepts are represented as "classes" with data fields ("attributes" describing the object) and associated procedures ("methods"). The objects interact to design applications and computer programs [2]. At its very basis, Object-Oriented Programming has an approach pointed toward problem solving, by requiring careful application of abstractions and subdividing problems into manageable pieces. Compared with procedural programming, the Object-Oriented code tends to break a program into vast numbers of small trivially verifiable pieces [3]. In the conventional model, the program is a list of tasks or subroutines to be performed, while the Object-Oriented program is a collection of interacting objects. The objects are capable of exchanging messages and processing data. Each object has a distinct role or responsibility as an independent apparatus.

The possible actions on these objects are defined by the methods and are closely associated with the object. Indeed, Object-Oriented data structures have the property of carrying their own operators or, at least, of inheriting them from a similar object or class.

Conversely, non-object-oriented programs may be seen as a "long" list of commands. In more complex programs, smaller sections of commands are inserted into functions or subroutines to perform a specific task. In this case, the designer tends to define the program's data as "global," that is, accessible from any part of the program. When code increases in dimension, permitting any procedure to modify any data, this might involve bugs with wide-ranging consequences.

On the other hand, Object-Oriented philosophy encourages the programmer to put data where there is no direct access by the rest of the program. In this approach, the data can be used only by the methods, grouped with the data, retrieving and modifying the controlled data. At this point, the object becomes the programming construct that combines data with a set of methods for accessing and managing those data. Using subroutines for examining or modifying certain kinds of data is a practice used in non-object-oriented modular programming, before of Object-Oriented Programming.

Through a set of designed functions, objects encapsulate and ensure that their data are used in an appropriate way. The methods typically include functions of check and guarantee integrity of the data types. Simple-to-use, and standardized methods are defined for performing particular operations on its data, concealing how those tasks are accomplished. In this way, the internal structure or methods of an object can be altered without requiring modifications to the rest of the program. This approach can also be used to propose standardized methods across different types of objects. As an example [4], several different types of objects might offer measurement procedures. Each of them might implement the measurement differently, reflecting the different kind of method it contains, but all the different measurement methods might be called in the same standardized manner from elsewhere in the program. This becomes particularly useful when more than one programmer is contributing to a project or when the goal is to reuse code between projects.

In the literature of the last 40 years [4], the main group of fundamental concepts for Object-Oriented Programming can be identified. The main concepts are: *Class, Object, Instance, Method, Message Passing, Inheritance, Abstraction, Encapsulation, Polymorphism, Decoupling, Dynamic Dispatch*, and *Open Recursion*. In the following, all these main concepts are reviewed [2–6].

Class: It defines the abstract characteristics of an entity (object), including the entity's characteristics (its attributes or fields) and its behavior (the things it can do, methods or operations). Classes provide modularity and structure in an object-oriented program. A class should naturally be decipherable to a nonprogrammer familiar with the problem domain, that is, the features of the class should have sense in the application context. Furthermore, its code should be quite self-contained (generally using encapsulation). Together, the properties and methods defined by a class are called members.

Object: It refers to a specific instance of a class. Each object has a structure similar to other objects of the class, but can have peculiar characteristics. An object can also call functions, or methods, specific to that object.

Instance variable: It is a variable defined in a class (i.e., a member variable), for which each object of the class has a separate copy, or instance.

Method: It is a subroutine (or procedure) associated with a class. Methods define the behavior to be demonstrated by instances of the associated class at program runtime. Methods have the special property at runtime of having access to data stored in an instance of the class (or class instance or class object or object) they are associated with and are thereby able to control the state of the instance.

Message Passing: It is a procedure call from one function to another. The message passing is the process by which an object sends data to another object or asks the other object to invoke a method.

Inheritance: It is a mechanism allowing data and behavior of one class to be included in or used as the basis for another class.

Abstraction: It is the action of defining classes to model aspects of reality using distinctions inherent to the domain of interest.

Encapsulation: It is a technique for designing classes and objects that restricts access to data and behavior by defining a limited set of messages that an object of that class can receive.

Polymorphism: It defines the ability of different classes to respond to the same message, implementing each method appropriately.

Decoupling: It refers to careful controls separating code modules from specific use cases, aimed at increasing code reusability.

Dynamic dispatch: It is the selection of the implementation technique for a polymorphic operation (method or function) to call at runtime.

Open recursion: It is a feature allowing a method body to invoke another method body within the same object via a special variable.

3.2.2 PATTERNS

Design is one of the main challenges in software development for measurement and test automation, and Object-Oriented Programming provides several specific methodologies. The design pattern is the most common methodology, as codified by [7]. In software design, the term "design patterns" is a general, repeatable solution to a frequently arising problem. From the operating viewpoint, it is a description or template for solving a problem in many different situations. This is not a finished design, and it cannot be transformed directly into source or machine code. In the application, the programmer implements the patterns that are most suitable in that case.

In particular, Object-Oriented design patterns typically represent relationships and interactions between classes or objects, and the final application of classes or objects is not specified. In functional programming languages, patterns are not applicable if they imply object-orientation or more generally a mutable state.

In general, a pattern has four essential elements [7]:

1. The **pattern name** is used to describe synthetically a design problem, its solutions, and effects. It makes it easier to think about designs and to communicate them and their trade-offs to others.
2. The **problem** describes when to use the pattern. It describes the problem addressed by the pattern and its context. It might define a particular design. Sometimes the problem will include a list of conditions that must be met before it makes sense to apply the pattern.
3. The **solution** describes the elements of the design, their relationships, tasks, and collaborations. It doesn't describe a particular concrete design or implementation, because a pattern is like a template to be applied in different circumstances. Conversely, the pattern offers a conceptual description of a design problem, and how a general arrangement of elements (classes and objects in our case) solves it.
4. The **consequences** are the results and trade-offs of the pattern application. Though often hidden when design decisions are described, consequences are critical for evaluating design alternatives and for understanding the costs and benefits of the pattern

usage. Consequences often deals with space and time trade-offs. They can also tackle language and implementation matters. Frequently, the consequences of a pattern include its impact on a system's flexibility, extensibility, reusability, or portability. Defining consequences helps their understanding and evaluation.

In software architecture, there are different levels of design. At a higher level, there are architectural patterns, describing an overall pattern followed by an entire system. At a lower level, there is the architecture specifically related to the purpose of the application. Still another level down, the architecture of the modules and their interconnections reside [8]. This is the domain of design patterns, packages, components, and classes.

There are many types of design patterns:

- The *strategy* pattern defines a family of algorithms, encapsulates them, and makes them interchangeable. It lets the algorithm vary independently from the using clients.
- *Computational design* patterns are related to key computation identification.
- *Execution* patterns refer to supporting application execution, comprising strategies in accomplishing streams of tasks and building blocks to support task synchronization.
- *Implementation strategy* patterns are suitable for implementing source code to support (1) program organization, and (2) the common data structures specific to parallel programming.
- *Structural design* patterns address the development of high-level structures of applications being developed.

Design patterns are composed of several sections. The main sections of a design pattern are Structure, Participants, and Collaboration. They describe a design motif—a prototypical microarchitecture—used by developers and adapted to their particular designs for solving the recurrent problem described by the design pattern. A microarchitecture is a set of constituents (e.g., classes and methods) and their relationships. A design pattern is used by software developers for introducing the prototypical microarchitecture of their designs, that is, the implemented design motif will have the same structure and organization of the chosen microarchitectures pattern. Another important feature of design patterns is to allow developers to communicate using well-known names for software interactions.

3.2.2.1 Patterns Classification and List

The design patterns can be classified by two criteria [7]: purpose and scope (Table 3.1). The first criterion reveals what the pattern really does. Consequently, design patterns are grouped into the following categories:

- *Creational* patterns are related to the process of object creation.
- *Structural* patterns are related to the structure of classes or objects.
- *Behavioral* patterns distinguish the modes how classes or objects interact and distribute responsibility.

The second scope-based criterion specifies whether the pattern applies primarily to classes or to objects [7]. Class patterns manage the relationships between classes and their subclasses. The relationships are created through inheritance, in order to be, static and fixed at compile-time. Object patterns manage object relationships, which are more dynamic because they change at run-time. Almost all patterns use inheritance to some extent, and for this, the only patterns labeled "class patterns" are those focusing on class associations. Most patterns are in the Object scope.

A description of these patterns is out of the scope of this book and can be found in [7].

A key issue with the general design patterns is to define their optimal use-area, that is, the particular types of software where they can be used with highest efficiency. It is also problematic to determine in which

Table 3.1. Classification of patterns

		Purpose		
		Creational	**Structural**	**Behavioral**
Scope	Class	Factory Method	Adapter	Interpreter Template Method
	Object	Abstract Factory Builder Prototype Singleton	Adapter Bridge Composite Decorator Façade Proxy	Chain of Responsibility Command Iterator Mediator Memento Flyweight Observer State Strategy Visitor

context or in which part of the system the patterns' attributes can be used [9]. A Domain Specific Design Pattern is a set of objects and components that form a highly encapsulated, cohesive partition with clear boundaries, which can be used in a specific software domain [10]. It would be a pattern that is in some way optimal for that particular domain [9]. In literature, different examples of domain-specific patterns can be found, such as user interface design patterns [11], information visualization [12], secure design [13], "secure usability" [14], Web design [15], and business model design [16].

3.2.3 ADVANTAGES IN MEASUREMENT APPLICATIONS

The development of an automatic measurement system is generally aimed at optimizing the final operation performance and cost ratio by simultaneously achieving the metrological target. In practice, structures of data acquisition systems have been quite standardized at hardware level, by mainly leaving the burden of software implementation for customizing the specific measurement application. For this reason, the need for reusability of existing software and flexibility of running applications is increased. In this context, the application to measurement systems of Object-Oriented paradigm carries significant advantages and satisfies the requirements of reusability and flexibility. When basing a measurement system on Object-Oriented principles, the actual world is modeled in terms of objects representing real-world entities as closely as possible [17]. In principle, all the steps of a test procedure and the components of a measurement system can be represented by an object (trigger, model, sensor, actuator, etc., as shown in Figure 3.1 [17]).

In literature, the application of the Object-Oriented paradigm is exploited for easily developing software for measurement and test applications under highly- and fast-varying requirements [18]. This paradigm associated to a partial design is a support structure where a measurement application can be organized and developed. An Object-Oriented system can be configured for satisfying a large set of measurement applications in a generic field for an industrial test division, an experimental laboratory, or a research center [18].

If the measurement system is conceived as a collection of active and cooperating entities (Figure 3.1), this approach improves the decomposition of the system in subsystems easier to be analyzed, decreases the dependencies between the various parts, and increases the system flexibility [17].

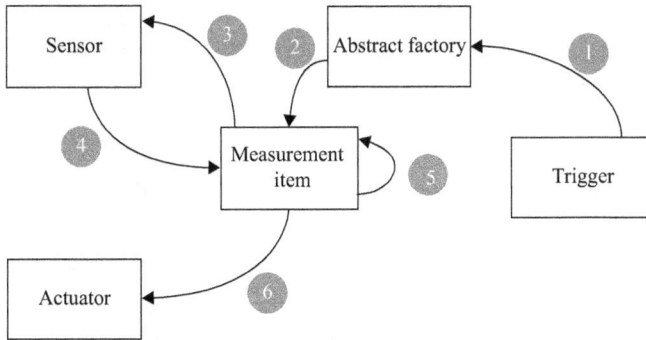

Figure 3.1. Architecture of a simple measurement and control system [17]: (1) a first event triggers the model (*abstract factory*), (2) the model produces a system software instance (*measurement item*), (3) the instance drives the sensor to carry out a measurement, (4) the sensor returns back the reading, (5) the model instance processes the data, and (6) drives the actuator suitably.

While test programs have been designed so far to solve specific problems with extremely limited capability to evolve, an Object-Oriented system for a given measurement domain, suitably conceived in order to be configurable for satisfying a wide range of requirements, could constitute a unified solution to drive all the existing and future park of measurement applications [18].

Although this approach leads to satisfy the need for reusability, flexibility, and quality of measurement software, the main drawback is the higher programming knowledge of final user or test engineer.

At commercial level, some products are provided for supporting the user in designing new tests (NI TestStand and Veristand, Director II Azimuth Systems, Activate Test Platform, see Chapter 1), by integrating software modules developed in other programming languages (C, C++, LabVIEW) [18]. However, these commercial platforms do not support the user in developing single software modules, and, as a result, standard development and reusability are intrinsically limited. Some examples of this approach are presented in [19, 20], where packages written in LabVIEW and LabWindows/CVI are designed for microscope control and for battery resistance measurement, respectively. In [21, 22], software development systems are proposed for measurement and control of data acquisition system. In [23], each task is best performed within a dedicated environment: LabVIEW, Matlab, custom simulators, and more.

At research level, the main advantages of Object-Oriented paradigm are used in the Object-Oriented Flexible Measurement System (OFMS) [24], and the Flexible Framework for Magnetic Measurements

(FFMM) developed by the University of Sannio in cooperation with CERN [18]. In these research projects, main emphasis is given to encapsulation, inheritance, flexible-construction, and multitask options. For a distributed measurement laboratory, an open architecture characterized by high reconfigurability, modularity, extendibility, and reusability is implemented by the development of Object-Oriented Programming [25]. The model, proposed in [26], represents the measurement systems by an Object-Oriented method, and in addition, highlights the limitations of the conventional function-oriented models. At Fermilab, the Object-Oriented software for magnetic measurements, EMS, allows easy reconfiguration and runtime modification, as well as various user interface, data acquisition, analysis, and data persistence components to be configured to form different measurement systems [27]. Again at CERN, Object-Oriented software for measurement and test applications allow the software quality in terms of flexibility, usability, and maintainability to be maximized [18].

3.3 ASPECT-ORIENTED PROGRAMMING

3.3.1 MOTIVATION AND BASIC CONCEPTS

Software systems can be viewed from multiple perspectives, referring even to different conceptual models. The model used to decompose a software system into modules has a radical impact on the modularization properties of the resulting software, and thus on its reusability and flexibility.

The decomposition criteria, in any way chosen, improve the modularization of some sets of functionalities widely needed in the system, called *concerns*, without affecting the others. The ideal situation would be not to choose the right criteria for modularization, but rather to have the possibility to decompose the system according to a set of criteria for each of the relevant concerns. Traditional Object-Oriented languages supporting only hierarchical decomposition do not provide developers with mechanisms powerful enough to accomplish this task: They lead the developer to a dominant decomposition that takes into accounts a certain number of "main" concerns (often called *primary concerns*). The others, often called *crosscutting concerns*, overlap on the hierarchy shaped by primary concerns.

In Figure 3.2, the example of the functions needed for the fault detection of some measuring devices is shown. These functions are spread over several objects, even if they have a common logic, which could give rise to a further component, the *Fault Detector*.

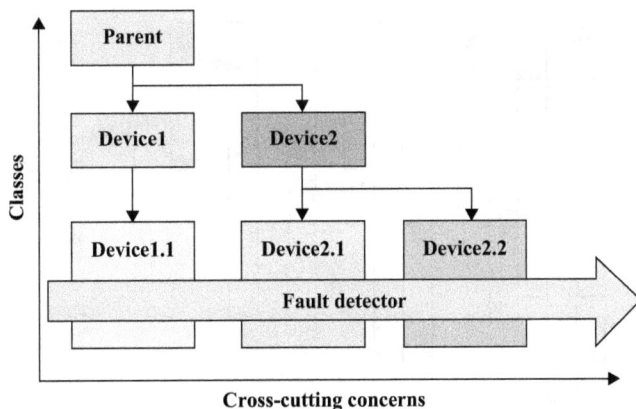

Figure 3.2. Example of a crosscuttings concern in measurement software: The fault detection [18].

The word "crosscutting" refers to the fact that such concerns are spread over the hierarchical modules of the dominant decomposition. The presence of crosscutting concerns in software systems has mainly two observable effects on the source code, called (a) "tangling," when in a single modularization unit there is code related to different concerns, and (b) "scattering," meaning that the implementation of a concern is spread across several different modularization units. Figure 3.3 highlights typical situations of (a) code scattering and (b) tangling.

In particular, in Figure 3.3, the vertical bars represent the modularization units and the different nuances of gray the diverse concerns. In the representation of Figure 3.3a, a highly scattered code has high level of distribution of a concern (rectangle with the same gray nuance) over different units (vertical bars). In the representation of Figure 3.3b, a highly tangled code has a high level of distribution of concerns (rectangles with a given gray nuance) over the same unit (vertical bar).

Crosscutting concerns are related to issues transversal to many modules, and thus cause the duplication of portions of code in several different modules, by negatively affecting maintainability and reusability. This means that at run-time each component must encapsulate its part of information related to the crosscutting concern even when it does not need it (thus wasting memory resources).

Crosscutting concerns can negatively affect the quality of even well modularized systems implemented by means of Object-Oriented [17], component-based [27], and agent-based techniques [28]. *Aspect-Oriented Programming (AOP)* [29] is an extension of the Object-Oriented paradigm

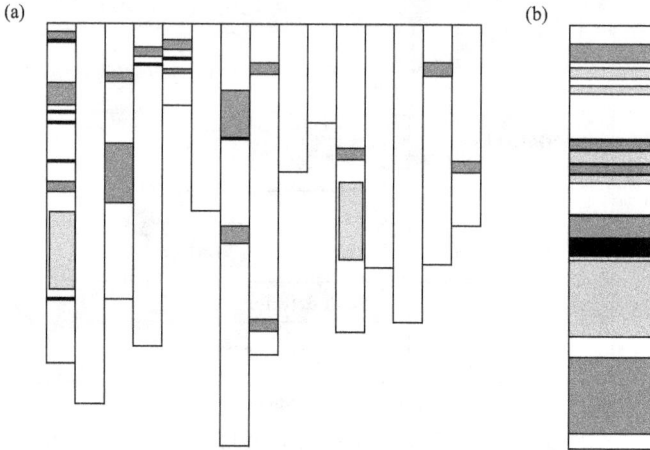

Figure 3.3. (a) Code scattering and (b) tangling.

that provides new specific constructs for improving the separation of concerns and supporting their crosscutting. AOP defines specific program units, the *aspects*, for specifying concerns separately, and rules for weaving them to produce the overall system to be run. The Aspect-Oriented architecture enforces, as much as possible, a centralized design, where the state of crosscutting concerns is maintained in dedicated modules, the aspects [29]. Each of them encapsulates a crosscutting concern for all the components involved with it. When a component does not need the feature of the crosscutting concern, the related data are not stored for it in the aspect, without corresponding memory waste.

As an example, tracing is aimed at recording report about code running for debugging or diagnosis purposes. The corresponding code is present approximately inside all the modules. This limits the reusability because this code is specific for the single module as well as for the application. According to the AOP approach, the aspect of tracing has to be removed from the functional concerns of the other modules and isolated in a specific module. Like a class of objects, an aspect introduces a new user-defined type into the system type hierarchy, with its own methods, fields, and static relationships to other types. Accordingly, an AOP system can be seen as composed by two parts: (a) a first one, consisting of traditional modularization units (e.g., classes, functions) and referred to as the base system, or *core concern*, and (b) a second one, consisting of aspects, encapsulating the crosscutting concerns involved in the system, and usually referred to as the *secondary concern*.

3.3.2   JOIN POINT MODEL

AOP provides twofold specific features for implementing crosscutting concerns in aspects: (a) *dynamic crosscutting features*, implementation of crosscutting concerns by modifying the runtime behavior of a program, and (b) *static crosscutting features*, modification of the static and structural properties of the system.

Dynamic crosscutting is implemented by using *advices* and *pointcuts*. An advice is a code fragment executed at specified points during the program runtime. The points in the dynamic control flow, where the advice code is executed, are called *join points*. A pointcut defines the events (such as a method call or execution, field get and set, exception handling and softening) triggering the execution of the associated advices. A pointcut is an expression pattern matched during execution to join points of interest. Each advice is associated to a pointcut by defining the join point(s) at which it must be applied. Advice code can be executed before, after, or around the intercepted join point. A *join point shadow* is the static counterpart, in the code, of a join point; equally, a join point is a particular execution of a join point shadow.

The aspects are inserted in the base program statically by a weaving process. The weaver is the component of an AOP programming language environment (such as AspectJ [30]) in charge of the weaving process. The weaver inserts instructions at join point shadows in order to execute the advice to be applied at the corresponding join points. The weaver may need to add runtime checks to the code inserted at a join point shadow in order to perform binding parameters, and other requested computations. The static crosscutting features of AOP implement crosscutting concerns by modifying the static structure of the system. An aspect has specific features (intertype declarations) for

- introducing new members (i.e., fields, methods, and constructors) to a class, or interface;
- changing or adding parents to classes or interfaces;
- extending a class from the subtype of the original super-class or implement a new interface.

3.3.3   SAMPLE IMPLEMENTATION

An example of a straightforward program (an AO version of the "Hello World" program implemented in AspectJ [31]), enlightening AO basic working, is reported in Figure 3.4. In the figure, the code of the class

```
// HelloWorld.java                          // GreetingsAspect.java
public class HelloWorld {                   public aspect GreetingsAspect {
                                                pointcut    callTellMessage()    :
    public static void tell(String         call(public         static        void
message) {                                  HelloWorld.tell*(..));
        System.out.println(message);
    }                                           before() : callTellMessage() {
                                                    System.out.println("Good
    public        static        void       morning!");
tellPerson(String    message,    String         }
name) {
        System.out.println(name + ", "          after() : callSayMessage() {
+ message);                                         System.out.println("Bye Bye!");
    }                                           }
}                                           }
```

Figure 3.4. An example of a straightforward AOP program implemented in AspectJ [31].

HelloWorld and the aspect *GreetingsAspect* are reported. The aspect defines a pointcut and two advices. The *callTellMessage()* captures calls to all public static methods with names that start with "tell." In the example, the pointcut captures the calls to the methods*tell(...)* and *tellPerson(...)* in the class *HelloWorld* taking any arguments. The two advices, one before and one after, associated to the pointcut *callTellMessage()* will cause the printing of the "Good morning!" and "Bye Bye!" text strings just before and after, respectively, each message printed by the methods *tell()* and *tellPerson()*.

3.3.4   ADVANTAGES IN MEASUREMENT APPLICATIONS

Software parts of an automatic measurement system are usually developed by exploiting Object-Oriented [17], component-based [27], and agent-based techniques [28]. They aim at organizing the software in modules, each one responsible for specified features, reducing their coupling, and maximizing their internal cohesion.

Crosscutting concerns can result in a significant quality loss even in well modularized automatic measurement systems designed by means of these techniques [32]. Typical crosscutting concerns related to issues transversal to many modules are the synchronization and fault detection tasks. They cause the duplication of portions of code in several different modules, and negatively affect its maintainability and reusability.

The good separation of concerns allowed by an AOP-based approach, conversely, influences positively software quality. It allows a high level of maintainability and reusability of the code: For each new element added to a measurement program, the code needed to handle a crosscutting concern for that new element is added to the hierarchy of the aspect modeling that concern.

This means that all the code related to the crosscutting concern is well modularized in aspects and subaspects of this hierarchy, therefore, the commonalities among the different subaspects can be well structured and factored out. As a consequence, the AOP design, with respect to the "traditional" OOP approach, exhibits a much more centralized design, decreasing code duplication and significantly growing the possibility of code reuse. This is because, a piece of software is more reusable if it has as few dependencies as possible in the context of its usage, and this is exactly the case of software modules implementing a single concern. In addition to that, an AOP-based architecture is usually not targeted at a specific system component, and the same architecture can be reused to handle a crosscutting concern in different components.

Properties like comprehensibility and maintainability are also improved by AOP. If each concern is implemented in a separate module that can be understood independently of the others, the system structure has no need to be understood as a whole in order to understand a part of it. The comprehension efforts can be then focused on the concern of interest while ignoring the other concerns.

In a software system, maintenance requires adding, removing, or changing a particular feature of the system. In order to change a feature, two main issues have to be addressed: (a) feature localization and (b) implementation update. When a concern is not clearly separated from the others (being spread over a large part of the code), its localization can become a very difficult and expensive task, because it is necessary to find all the pieces of code contributing to that particular concern. Similar observations apply to the addition or the elimination of concerns. In the first case, all the locations where code has to be inserted have to be found, while in the latter case, all the locations where code has to be removed. Moreover, adequate testing to make sure that the changes in all scattered fragments do not introduce faults or subtle side effects is highly nontrivial. A good separation of concerns is one of the focal points of an AOP design; therefore, in systems developed according to such an approach, the maintainability is highly increased.

In the context of software engineering, separation of concerns is strictly related to composition and decomposition mechanisms, constructs and patterns as supported or enforced by programming languages and frameworks. Software composition and decomposition tasks lead to the partitioning of a software system into smaller parts that are less complex (decomposition task) and the assembly of software systems in terms of these smaller parts to build up the entire system (the composition task). In the past, most software systems were developed under the assumption of interaction with a static external environment with no (or very-limited) capability to evolve. Under such an assumption, developers were considering requirements stable and evolution as a problem to avoid. This assumption is not suitable in the field of software for measurement applications when high capability to evolve is needed in order to meet variable and changing requirements.

As a consequence, software for measurement applications greatly benefits from features that make system composition and decomposition more flexible and dynamic, with run-time binding in which the relationship among elements is established at run-time. This is another of the advantageous features offered by AOP.

Finally, when working with an AOP design, it is possible to focus on every single concern, ideally without caring about the others. This means that domain and technology experts can be assigned to each particular portion of the system, each one addressing a well-defined concern. This also means that tasks can be divided into smaller and more specific pieces that can be concurrently carried out by different people. With a good separation of concerns, the required coordination is minimized, because only people working on the same concern need to be coordinated.

REFERENCES

[1] Odei Bempong, B. 2013. "The Cognitive Programming Paradigm—The Next Programming Structure." *American Journal of Software Engineering and Applications* 2, no. 2, pp. 54–67. doi: http://dx.doi.org/10.11648/j.ajsea.20130202.15

[2] 2007. "Object Oriented Programming, OOP." *TechTerms*. http://www.techterms.com/definition/oop.

[3] 2014. "Object_Oriented_Programming." *WikiBooks*. http://en.wikibooks.org/wiki/Object_Oriented_Programming.

[4] Armstrong, D.J. February 2006. "The Quarks of Object-Oriented Development." *Communications of the ACM*, 49, no. 2, pp. 123–8. doi: http://dx.doi.org/10.1145/1113034.1113040.

[5] Slagell, M. 2008. "Methods." http://www.rubyist.net/~slagell/ruby/methods.html

[6] AbdelGawad, M. 2012. *NOOP: A Mathematical Model of Object-Oriented Programming*, Doctoral Thesis, Rice University's digital scholarship archive, Rice University, Houston Texas. http://scholarship.rice.edu/handle/1911/70199

[7] Gamma, E., R. Helm, R. Johnson, and J. Vlissides. 1994. *Design Patterns: Elements of Reusable Object Oriented Software*, http://www.uml.org.cn/c++/pdf/DesignPatterns.pdf

[8] Martin, R.C. 2000. *Design Principles and Design Patterns*. www.objectmentor.com

[9] Gustavsson, R., J. Ala-Kurikka, and S. Rulli. (2002). *Domain Specific Design Patterns, A Report in the Course Object-Oriented Programming Advanced Course*. Mälardalen University, Västerås, Sweden. http://www.idt.mdh.se/kurser/cd5130/msg/2002lp3/download/CD5130%20VT02%20DomainSpecificPatterns.pdf

[10] 1998. "Prototype Design Pattern." *Sourcemaking*. http://sourcemaking.com/design_patterns/prototype.

[11] Laakso, S.A. *Collection of User Interface Design Patterns*. University of Helsinki, Department of Computer Science, (January 31, 2008). http://www.cs.helsinki.fi/u/salaakso/patterns/

[12] Heer, J., and M. Agrawala. 2006. "Software Design Patterns for Information Visualization." *IEEE Transactions on Visualization and Computer Graphics* 12, no. 5, pp. 853–60. doi: http://dx.doi.org/10.1109/tvcg.2006.178

[13] Dougherty, C., K. Sayre, R.C. Seacord, D. Svoboda, and K. Togashi. 2009. "Secure Design Patterns," *Software Engineering Institute, Carnegie Mellon University*. http://www.cert.org/archive/pdf/09tr010.pdf

[14] Garfinkel, S.L. 2005. "Design Principles and Patterns for Computer Systems that Are Simultaneously Secure and Usable." *Simson Garfinkel's PhD thesis*. http://simson.net/thesis/

[15] 2012. "Yahoo! Design Pattern Library." *Yahoo Developer Network*. http://developer.yahoo.com/ypatterns/

[16] 2010. "How to Design Your Business Model as a Lean Startup." *The Methodologist*.http://torgronsund.com/2010/01/06/lean-startup-business-model-pattern/

[17] Bosch, J. 1999. "Design of an Object-Oriented Framework for Measurement Systems." In *Domain-Specific Application Frameworks*, eds. M. Fayad, D. Schmidt, and R. Johnson, pp. 177–205. New York, NY: John Wiley. ISBN 0-471-33280-1.

[18] Arpaia, P., M. Buzio, L. Fiscarelli, and V. Inglese. November 2012. "A Software Framework for Developing Measurement Applications Under Variable Requirements." *Review of Scientific Instruments* 83, no 11, 115103. doi: http://dx.doi.org/10.1063/1.4764664

[19] Langer, D., M.V. Hoff, A.J. Keller, C. Nagaraja, O.A. Pfäffli, M. Göldi, H. Kasper, and F. Helmchen. February 2013. "Helioscan: A Software Framework For Controlling In Vivo Microscopy Setups With High Hardware Flexibility, Functional Diversity And Extendibility." *Journal of Neuroscience Methods* 215, no. 1, pp. 38–52. doi: http://dx.doi.org/10.1016/j.jneumeth.2013.02.006

[20] Yongjie, F. August 2011. "Design of the Battery Resistance Measurement System." In *Electronic Measurement & Instruments (ICEMI)* 2011, 2, pp. 240–243. Chengdu, China: IEEE.

[21] Chen, Y. March 2009. "Electric Quantity Test System of Unified Power Flow Controller Model on LabWindows/CVI." *APPEEC 2009*, pp. 1–4. Wuhan, China: IEEE.

[22] Vişan, D.A., and I.B. Cioc. May 2010. "Virtual Instrumentation Application for Vibration Analysis in Electrical Equipments Testing." *ISSE 2010*, pp. 216–219. Warsaw, Poland: IEEE.

[23] Deshmukh, A., F. Ponci, A. Monti, L. Cristaldi, R. Ottoboni, M. Riva, and M. Lazzaroni. April 24–27, 2006. "Multi Agent Systems: An Example of Dynamic Reconfiguration." *IMTC 2006—Instrumentation and Measurement Technology Conference*. Sorrento, Italy: IEEE.

[24] Shen, X., X. Song, and J. Chen. August 16–19, 2009. "Implementation and Evaluation of Object-Oriented Flexible Measurement System." *Electronic Measurement & Instruments, ICEMI '09. 9th International Conference*. Beijing, China: IEEE.

[25] Arpaia, P., F. Cennamo, P. Daponte, and M. Savastano. June 4–6, 1996. "A Distributed Laboratory Based on Object-Oriented Measurement Systems." *IEEE Instrumentation and Measurement Technology Conference*, pp. 27–32. Brussels, Belgium: IEEE.

[26] Yang, Q., and C. Butler. February 1998. "An Object-Oriented Model of Measurement Systems." *IEEE Transactions on Instrumentation and Measurement* 47, no. 1, pp. 104–07.

[27] Nogiec, J.M., J. DiMarco, S. Kotelnikov, K. Trombly-Freytag, D. Walbridge, and M. Tartaglia. June 2006. "A Configurable Component-Based Software System for Magnetic Field Measurements." *IEEE Transactions on Applied Superconductivity* 16, no. 2, 1382–85.

[28] Jennings, N.R. 1999. "Agent-Based Computing: Promises and Perils." *Proceedings of the Sixteenth International Joint Conference on Artificial Intelligence (IJCAI)*, Vol. 2, pp. 1429–36. Stockholm, Sweden: Thomas Dean.

[29] Kiczales, G., J. Lampin, A. Mendhekar, C. Maeda, C. Videira Lopes, J.M. Loingtier, and J. Irwin. 1997. "Aspect-Oriented Programming." In *Proceedings of the 11th European Conference on Object-Oriented Programming (ECOOP)*, Vol. 1241, pp. 220–42. Jyväskylä, Finland: Springer-Verlag,

[30] Kiczales, G., E. Hilsdale, J. Hugunin, M. Kersten, J. Palm, and W.G. Griswold. 2001. "An overview of AspectJ." In *Proceedings of the 15th European Conference on Object-Oriented Programming (ECOOP 01)*, Vol. 2072, pp. 220–42. Budapest, Hungary: Springer-Verlag.

[31] 2014. "AspectJ." *Eclipse Foundation*. http://www.eclipse.org/aspectj/

[32] Pfister, C., and C. Szyperski. 1996. "Why Objects are Not Enough." In *Proceedings First International Component Users Conference (CUC 96)*, 3, pp 141–7. Munich, Germany: SIGS Book.

PART II

METHODOLOGY

CHAPTER 4

A FLEXIBLE SOFTWARE FRAMEWORK FOR MEASUREMENT APPLICATIONS

4.1 OVERVIEW

This chapter presents the design of a software framework for automatic measurement applications, based on Object-Oriented and Aspect-Oriented Programming. The objective of the framework is to make the development of new measurement programs simple and cost-effective, simultaneously allowing for easy modification and extension of existing test software. First, the paradigm of the framework is introduced, by highlighting the basic ideas leading to its conception and design, as well as its architecture at the structural and functional level. Then, the main components of the framework are described, by firstly introducing their corresponding state of the art, leading concepts, and architecture with their main modules. The review starts with the *Fault Detector*, aimed at identifying and locating failures and faults transparently to the user. Fault detection is a crosscutting concern; therefore, the design is led by an aspect-oriented approach. The reader will also find a meaningful practical example, based on actual on-field experience, of the design of an Aspect-Oriented Programming (AOP)-based component for measurement applications. Then, the *Synchronizer*, aimed at coordinating measurement tasks with well-defined high-level software events (e.g., start and stop, or device events) is presented. In particular, the reader will understand how the use of a Petri net (PN) modeling the execution path allows software synchronization to be abstracted above the code level, by leaving the test engineer to work at a more intuitive level. The review continues with the Measurement Domain

Specific Language (MDSL), aimed at defining specifications for complete, easy-to-understand, -reuse, and -maintain applications efficiently and quickly by means of a script. The related sections highlight the design for abstracting key concepts of the domain allowing the test engineer to write more concise and higher-level programs with natural language-like sentences in a shorter time without being a skilled programmer. Finally, the *Automatic Generator of User Interface*, aimed at separating the user interfaces easily from the application logic for enhancing the flexibility and reusability of the software, is illustrated. The corresponding sections highlight how a model-based approach, the Model-View-Interactor Paradigm, allows the designer to focus on the "interaction" typical in a software framework for measurement applications (SFMA) between the final user and the automatic measurement system.

4.2 FRAMEWORK PARADIGM

Chapter 2 highlighted how software frameworks are cohesive artifacts of design and implementation. This section defines a framework specific for measurement applications and highlights the specific relationships with its environment. The environment includes use-relationship-based (black-box) clients, inheritance-based (white-box) extension clients, and further classes the framework exploits [1, 2]. A black-box framework is expected to work out of the box: A client (object) can use the framework by instantiating classes and composing the instances to suit its needs. A white-box framework requires clients to supply new subclasses first, before objects can be created and composed. The approach presented here combines both the characteristics, by providing both readily usable classes and abstract classes to be subclassed in order to provide application-specific classes [3].

Another very general and very powerful principle exploited in the framework design is the separation of concerns presented in Chapter 3. Specifically for the framework, Aspect-Oriented Programming has the ability to identify, describe, and handle peculiar and critical facets of a measurement system separately. Concerns are always related to a goal a measurement field stakeholder wants to achieve, or to anticipations or expectations he or she has on the final measurement [4]. Aspect-Oriented approach provides a technique that is able to achieve a higher degree of separation of concerns vital for achieving flexibility.

In the following, (a) the *basic ideas*, (b) the *architecture*, and (c) the *design* of the framework are presented.

4.2.1 BASIC IDEAS

From prior considerations, in the following, the basic ideas underlying the conception of a generic framework for measurement applications are summarized:

1. A group of interfaces and abstract classes represents a white-box layer defining the high-level structure of a measurement framework for generating new parts of the framework itself; this allows potentiality and flexibility to be extended.
2. A group of modules represents a black-box layer, allowing both module reusability and ease of use to be attained, even by test engineers without knowledge of internal framework mechanisms.
3. Aspect-Oriented Programming improves the reusability and maintainability of a software framework [5]: In large projects, several concepts are transversal to many modules (cross-cutting concerns); they are extrapolated from the native units and implemented in separated modules (aspects), in order to improve the system modularity and enhance maintainability.
4. A library of reusable modules is built incrementally during the start-up of the framework up to a "saturation" condition inside an application domain, allowing progressively further requirements in the same domain to be satisfied by a limited effort.
5. A suitable definition of the code structure allows standard modules to be developed: Such modules represent a sound basis for a library both for implementing new components and for extending old ones.

According to these ideas, the working principle of a software framework for measurement applications (SFMA) is derived (Figure 4.1).

The test engineer produces a description of the measurement application, the *User Script*, whose syntactic correctness is verified by a *Script Checker*. Then, from the User Script, the *Builder* assembles the *Measurement Program*, according to the architecture of the *Scheme* by picking up suitable modules from the *Software Module Library*. If some modules are not available in the library, a template is provided to the user (administrator user) in order to implement them according to a suitable predisposed structure. Once debugged and tested, the *Measurement Program* will be stored in the *Database* (DB) in order to be reused.

According to the analysis of typical measurement use-case tests (or procedures), the generic *User Script* is organized into the following phases:

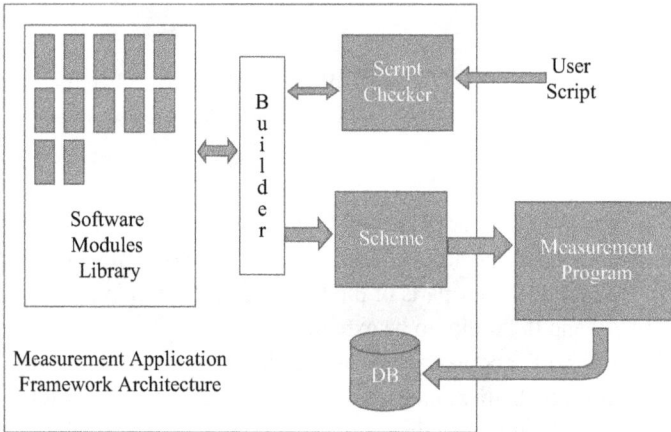

Figure 4.1. Working principle of a software framework for measurement applications.

- Definition of the measurement components;
- Specification of mechanical and electrical connections;
- Definition of dynamic parameters, that is, configurable during run-time of the Measurement Program;
- Component checking (fault detection);
- Storing of measurement conditions;
- Configuration of measurement devices;
- Description of the measurement procedure;
- Preliminary data analysis;
- Data saving.

4.2.2 ARCHITECTURE

The framework for measurement applications has a layered architecture (Figure 4.2), where each layer has an internal Object-Oriented organization. In this way, the objects interact only inside the layer horizontally, among entities of the same level. The layer features are realized by the corresponding objects, using in turn the capabilities of the upper level only through a suitable interface, such is typical in layered systems.

In the *Basic Service Layer*, all the services needed to implement the high-level logic of the framework are collected. Subcomponents for environment abstraction, memory management, error handling, file-system abstraction, and processes and threads handling are included. Furthermore, abstract communication services to higher-layer components allow

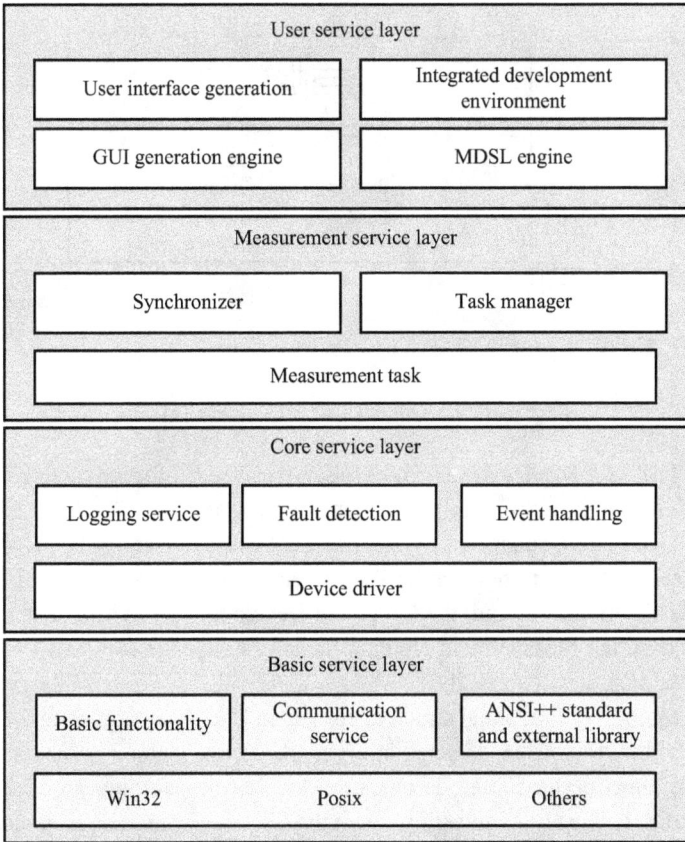

Figure 4.2. Layered architecture model of the SFMA.

data to be extracted from actual devices and external interfaces, by permitting communication mechanisms to be changed without significant performance loss.

In the *Core Service Layer*, several packages exposing main functionalities related to components (in particular measurement devices), event handling infrastructure, fault detection, and logging, are included.

In the *Measurement Service Layer*, a minimal but extensible infrastructure for managing a measurement procedure is offered by: (a) the *Test Manager*, encapsulating the user script for its execution in a controlled environment, and providing the user with core services for implementing the measurement process; (b) the *Measurement Tasks*, created to manage specific jobs of the instrumentation; and (c) the *Synchronizer*, for synchronizing the Measurement Tasks with well-defined high-level software events (e.g., start and stop, or device events).

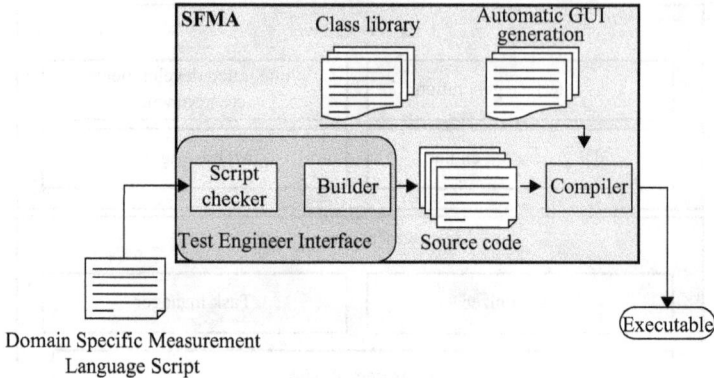

Figure 4.3. Architecture of a SFMA [6].

The *User Service Layer* provides services needed for interacting with the framework users mainly by means of two modules: The User Interface Generation, supported by the graphical user interface (GUI) engine, devoted to the test engineer for easily generating professional GUIs for the application user; and, the Integrated Development Environment, supported by the Measurement Domain Specific Language (MDSL) Engine for defining the script easily.

Multilayer functional structure corresponds to the operating architecture of the framework depicted in Figure 4.3. This operating architecture is the direct derivation at the operation level of the working principle in Figure 4.1. The test engineer produces the *User Script* in MDSL by means of a suitable interface based on the Integrated Development Environment. The MDSL also allows the final user interface to be specified for the automatic GUI generation.

4.2.3 DESIGN

The UML model of the framework kernel is shown in Figure 4.4, where its main components and the relations among them are highlighted. The *TestManager* organizes the test by knowing the device under test (*UnitUnderTest*), the measurands (*Quantity*), the measurement configuration, and the measurement procedure (written in MDSL within the User Script). The *TestManager* has an association with the *Devices* (software representation of the measurement devices). Among *Devices*, the PC can control remotely the *VirtualDevices* through a *CommunicationBus*. The system also represents the sensors and transducers in dedicated class hierarchies. The *Synchronizer* and the *FaultDetector* are critical modules for a

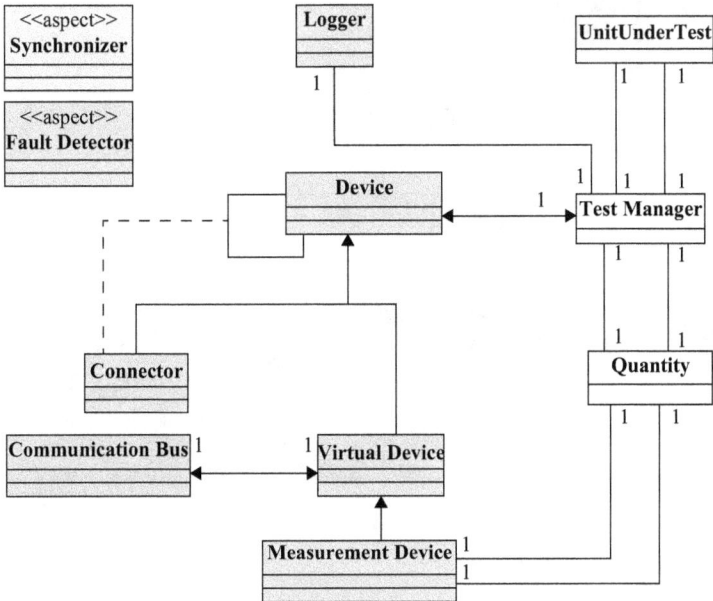

Figure 4.4. The UML model of the framework kernel.

measurement application: The former allows the measurement algorithm timing at the software level, while the latter fosters the identification and the location of failures and faults transparently to the user.

Both such features are transversal to several functional units (cross-cutting concerns): The synchronization policy involves all the measurement devices and all the test procedures, while the fault detection is a fundamental part of all the devices, as well as of the measurement system as a whole. The *Synchronizer* and the *Fault Detector* can be, therefore, encapsulated in aspects according to the AOP approach. The synchronization policy and fault management strategy can be extrapolated from the single modules and handled separately. In this way, future changes related to these topics will affect only the *Synchronizer* and the *Fault Detector* modules, without involving all the related classes, directly or indirectly, to the fault or synchronization events.

The architecture is further detailed in Figure 4.5, where its multilayered structure is shown (for the sake of simplicity, the User Service Layer is not represented). In this structure, the abstraction level grows bottom-up from the hardware toward to the user discretely layer by layer. Single layers have an Object-Oriented internal organization corresponding to the same abstraction level. Thus, the interaction among objects takes place horizontally, among entities of the same abstraction level. The features of the

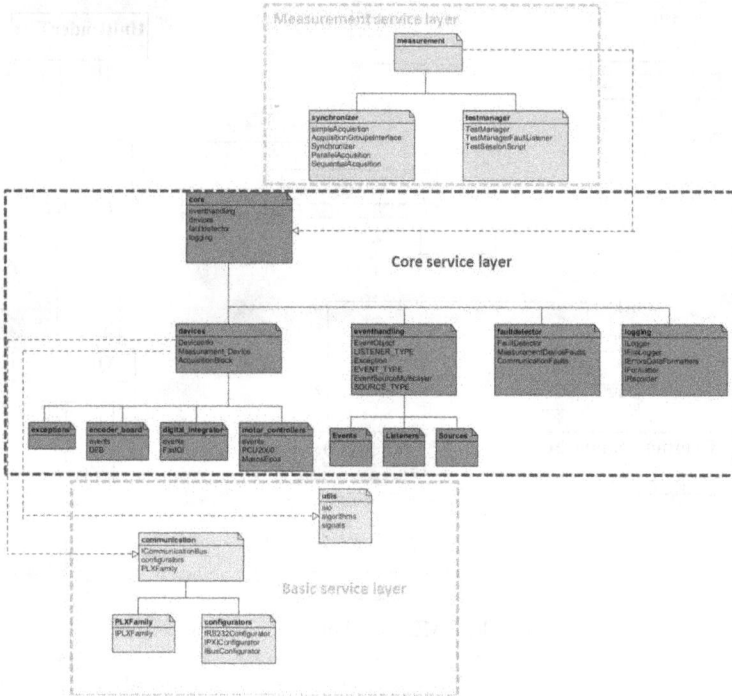

Figure 4.5. UML diagram of the multilayered architecture of the framework (for the sake of simplicity the *User service* layer is not reported).

level *i* are realized by the objects of the corresponding layer, by exploiting the capabilities offered by level *i*-1, through a suitably defined interface, as typically happens in layered systems. However, this layered structure is not rigid. Even though it would be better to use, while programming in a level, only functions implemented in the next lower layer, the user is allowed to call the functionalities of all the underlying levels, in order to improve the design flexibility.

In the bottom layer (*Basic service*), all the lower-level basic services needed to implement higher-level logic are placed. This layer includes subcomponents for environment abstraction, memory management, error handling, and file system abstraction, as well as processes and threads handling. It also defines abstract communication services to higher-level components within the measurement framework layers, in order to extract data from actual devices and external interfaces, by allowing the exchangeability of the communication mechanisms without incurring performance penalties. The interface *ICommunicationBus* is used to send and receive data to and from components in an abstract way. Concrete implementations of such interface are required to handle specific communication devices.

The middle layer in Figure 4.5 (*Core Service*) includes several packages exposing main functionalities related to components (in particular measurement devices), event handling infrastructure, fault detection, and logging. In the design of the event handling architecture, a variant of the Observer design pattern [4] is used in order to keep synchronized the state of cooperating components (e.g., *VirtualDevices*). The Observer enables one-way propagation of changes: One publisher notifies any number of subscribers about changes of its state, thus providing a form of loosely coupled signaling from publisher to subscribers. In Figures 4.6 and 4.7, the main architecture and the related infrastructure, respectively, are shown.

Logging facilities are also provided at this level of the architecture (Figure 4.5, *Core Service*). The class *Logger* handles the storage of configuration and measurement data, as well as system warnings and exceptions. The logger architecture is detailed in Figure 4.8. Data can be stored in a text or binary file. In any case, the final destination of the logged messages has to be kept decoupled with the format of the messages themselves. With this aim, two different responsibilities arise: logged message formatting and logged message recording. The formatter does not take care about where the message is recorded, and the recorder does not care about the format of the message. Therefore, the class *Logger* implements the design pattern Strategy [4]: The concrete logger can be configured with the right formatter and the right recorder keeping them decoupled.

Based on the services provided at this level of architecture, the *Measurement service* layer (top of Figure 4.5) implements a minimal but extensible infrastructure, based on the class *TestManager*, in order to handle and perform measurement sessions. For the measurement layer, two main features are needed:

- A test session director, the *TestManager*, encapsulating the user script and executing it in a controlled environment. Within the user scripts, core services are made available to the user in order to implement its measurement process.
- The capability of creating groups of data acquisition tasks (measurement tasks) to be synchronized with well-defined events (e.g., start and stop, or device events). In the measurement framework, the component realizing this high-level software synchronization of data acquisition is the *Synchronizer* (Figure 4.5).

In the following sections, further details about the most important components of the framework architecture are presented.

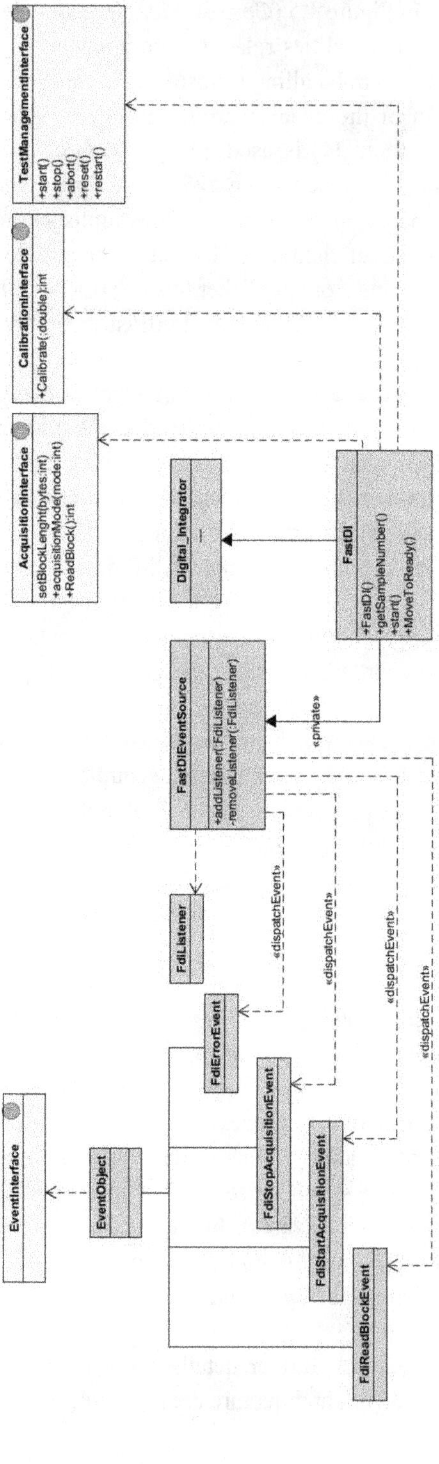

Figure 4.6. Measurement framework event handling architecture.

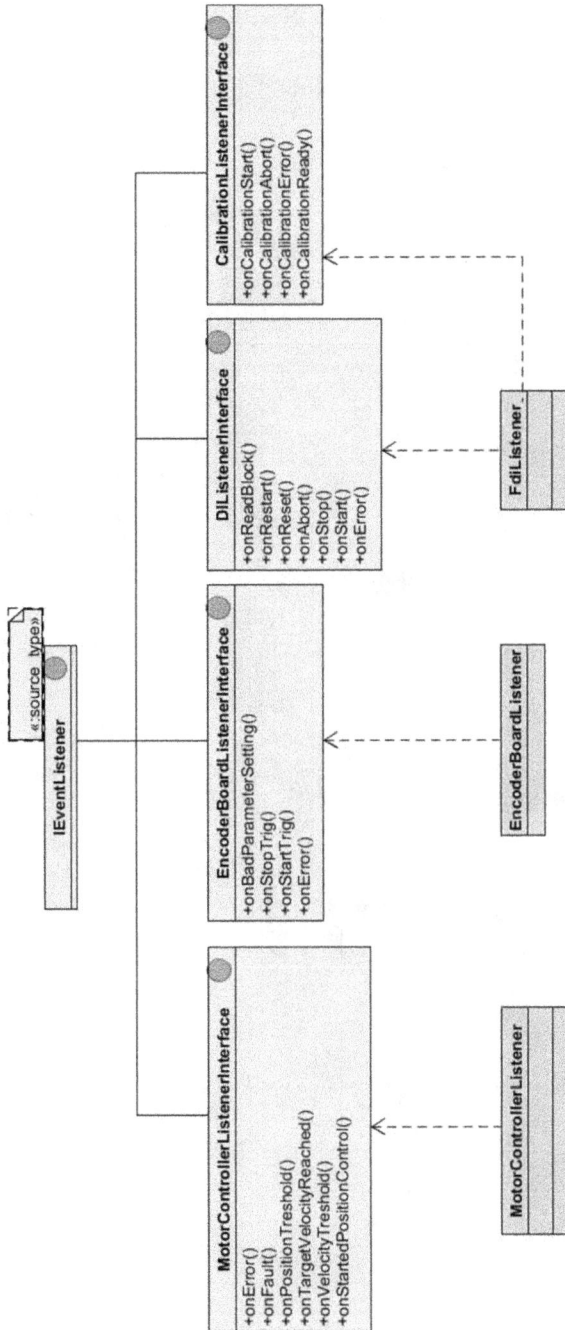

Figure 4.7. Measurement framework actions and listeners infrastructure.

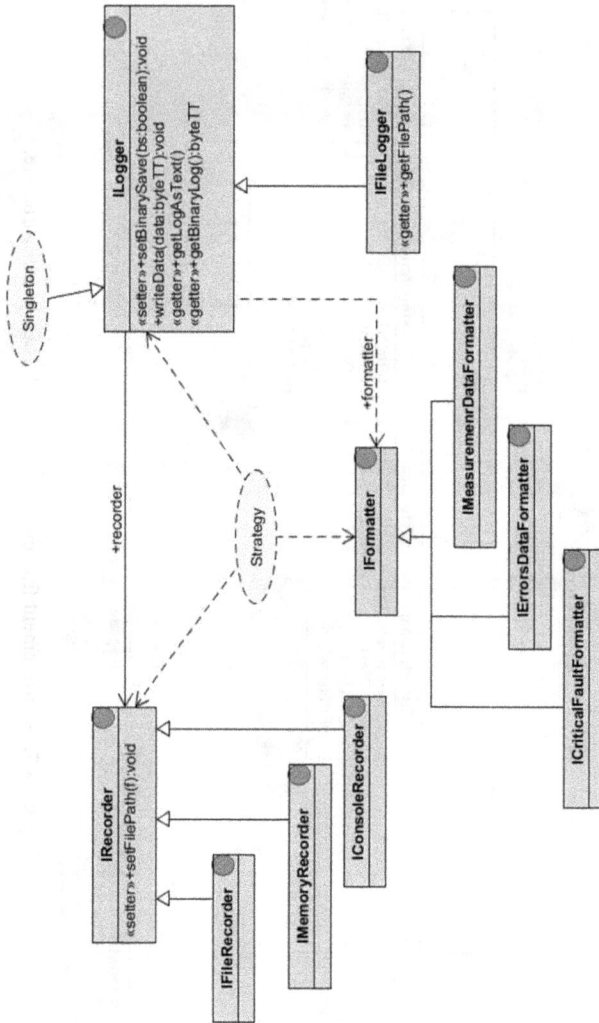

Figure 4.8. Logger architecture of the framework.

4.3 FAULT DETECTOR

The *Fault Detector* is one of most critical modules in the design of a framework for measurement application. This module promotes the identification and location of failures and faults transparently to the user. As highlighted in previous sections, the fault detection is a transversal service and a fundamental part of all the devices, as well as of a measurement system as a whole. In AOP approach, the *Fault Detector* is encapsulated in aspects. In this way, fault management strategy can be extrapolated from the single modules and handled separately, and future changes related to this topic will affect only the *Fault Detector*, without involving all the related classes, directly or indirectly, to the fault events.

In the following, an AOP-based approach to the development of software for fault detection in automatic measurement systems is described. In particular, (a) a short introduction to *fault detection in measurement automation*, (b) the *basic ideas* underlying the design, and (c) the *architecture* of the *Fault Detector* are highlighted.

4.3.1 FAULT DETECTION IN MEASUREMENT AUTOMATION

In measurement and test automation involving several instruments, one of the key design issues of smart systems is the capability of assuring a proper termination to the test process in case of an anomaly. Furthermore, a prompt detection of the anomaly improves the maintenance effectiveness of the measurement system, by highlighting possible sources of failures before their actual occurrence. With these goals in mind, a suitable software strategy for automatic fault detection turns out to be an adequate reaction to anomalous working, both at the software and hardware level [7, 8]. Devices provide information about their status continuously and, in case of abnormal working, a fault condition is pointed out. Compared to a fully hardware implementation, it has the advantage of higher flexibility and cost effectiveness. Today, it is a widely used technique, and the emergence in the area of application for cost-effective dependable systems will further increase its importance [9]. Thus, the software implementation of a fault detector also affects the overall system quality, in particular in terms of maintainability and reusability.

The analysis of the state-of-the-art automatic measurement systems highlights that fault detection is usually scattered all over different software components, and, in particular, mainly over the devices hierarchy [10]. This means that often the concrete classes of virtual devices contain duplicated code for fault detection, thus making harder their comprehension, testing, and maintenance. The Aspect-Oriented Programming shown here is capable of overcoming the drawbacks arising from the intrinsic crosscutting nature of the fault detection components. The crosscutting concerns related to fault detection of a large measurement software project are separated and handled better by encapsulating them into aspects. In this way, the reusability of system modules is improved.

4.3.2 BASIC IDEAS

Most common faults in an automatic measurement system can be classified according to (a) *their sources,* and (b) the *synchronization of the related handling operations* [10].

According to *their sources,* faults can be classified as arising from

- *Hardware devices*, when in a faulty internal state due to hardware anomaly or to an external condition. A device's internal fault detection can scale from very basic internal information to very complex routines forcing the device in different states. Correspondingly, concrete aspects of the fault detection subsystem must be capable of intercepting relevant changes in the device status, decoding them, and broadcasting high-level faults description to the interested components;
- *The measurement environment*, when affected by external or internal alterations. A typical example is an anomalous alteration of temperature or electromagnetic noise arising from the system under test and affecting the instrument's behavior;
- *Software components*, when in any nonconsistent state, owing to an incorrect use violating pre- and post-conditions, or to the presence of unresolved or undiscovered bugs.

According to *the synchronization of the related handling operations,* faults can be classified as in the following:

- *Synchronous*, when an anomalous operation is attempted. In this case, the following policies can be applied, according to the criticality level and the kind of the fault:
  - *k-times retry*: Some operations are retried until the device goes back in a consistent state successfully, without any performance constraint on the operation. As an example, an initialization reset tried several times during a slow start up of a multimeter.
  - *Multicast warning and continue*: For operations commanded under wrong conditions, the corresponding requests can be

ignored by issuing only a notification warning. As an example, a digital scope is triggered when previous data digitization is not ended, or when a stop or an abort is issued on an already stopped instrument.

o *Multicast fault and deny operation:* For operations not to be executed when specified faults occur. In this case, the operation is denied and the fault information is sent to pertinent components in order to be properly handled. This is the case when a multimeter has a fusible broken fusible during a measurement of an extremely high current value and it is not capable of carrying out the measurement.

o *Multicast an immediate shutdown request and deny operation:* In the most critical situations, a fault in a risky device during a critical operation should be blocked at the lowest level. Moreover, the system as a whole is to be shutdown gracefully as fast as possible, thus a high-priority request of system shutdown is sent to the fault handler component. This is the case when an overheating occurs in a shunt resistor owing to a wrong value of current setting. These faults are handled suitably by wrapping operations through concrete pointcuts, surrounded by advices defined by abstract aspects of the fault detector component.

• *Asynchronous:* Faults generated by hardware or environment anomalies occurred in whatever instant not synchronized with the measurement operations. The related detection is based on field access point-cut expressions bound to the decoding logic used to detect changes in the status of devices.

4.3.3 ARCHITECTURE

The architecture of the *Fault Detector* is based on two main subsystems:

• A fault detection subsystem, designed for
  o monitoring the "health" status of the measurement devices;
  o catching software faults such as stack overflow, live-lock, dead-lock, and application-defined faults, as soon as they occur.
• A fault notification subsystem, responsible for [10]
  o constantly receiving the sequence of occurring faults in real time from all the system components;
  o storing the detection history and providing access to other components or to external humans in order to react to faulty events adequately.

These two subsystems exploit three key software components: (*a*) a *FaultDetector* aspect hierarchy (Figure 4.9), allowing the code related to the fault detection logic to be removed from the modules implementing the virtual devices; (*b*) *FaultDecoder* tables, needed by concrete aspects for decoding status representation specific of concrete *VirtualDevices*; and (*c*) *FaultListeners* in order to dynamically bind (obliviously) components responsible for the fault management to the ones acting as fault sources.

The aspects in the *FaultDetector* hierarchy intercept faults by means of the *FaultDecoder* classes. The decoders are capable of handling groups of similar devices and knowing internal state structure and encoding. They provide the aspect logic by *FaultTable* instances encapsulating fault information to be sent to the interested components through a *FaultNotifier* layer.

In Figure 4.9, the *FaultDetector* hierarchy is depicted, by highlighting the static relationships among *VirtualDevice* classes, *FaultDetector* aspects, and some concrete virtual devices. The figure shows the role played by the *FaultDecoder* and *FaultTable* for a generic *VirtualDevice1*. Encoded fault information is extracted from the device *VirtualDevice1* by context interception and is decoded by a concrete *VirtualDevice1_Fault-Decoder*.

The decoded information is then provided to the *VirtualDevice1_FaultDetector*, responsible for enforcing fault management policies according to the fault kind. Moreover, the *VirtualDevice1_FaultDetector* sends the fault data to the interested software components.

The *FaultDetector* is responsible for defining high-level point-cuts capturing relevant operations affecting the status of devices. In the measurement system, the *VirtualDevice* hierarchy [10] models and organizes all the physical devices involved in the measurement process. Each device has an internal status; modifications to such status are captured by means of concrete sub-aspects executing the logic needed to decode it, as well as detecting if and where a device notified an internal fault. In each *FaultDetector* sub-aspect, associated to main devices categories, the mapping logic toward concrete devices classes belonging to the same family is defined and the common behaviors can be factorized, as needed. The coarseness of the mapping among aspects and concrete devices allows a very flexible reuse of fault detection logic for similar devices by encapsulating it in few modules (instead of spreading it all over the device classes).

Figure 4.10 depicts the different levels of fault interceptions, according to the fault types. The bottom level takes care about very specific issues and features of concrete devices to encapsulate in dedicated subaspects. At the middle level, concrete aspects, by using decoders, perform continuous monitoring of devices' status.

Slice Class

+Virtual boot checkDeviceStatus()
+Virtual int decodeError()
+Virtual void addListener()
+Virtual void removeListener()

<<aspect>>FaultDetector

- _ mdevs: Vector Virtual_Device*

+<<pointcut>>+devices()
+<<advice>>+devices() : Slice Class
+void addToMonitor(in ffmm::core::devices::Virtual_Device* m)
+void removeFromMonitor(in ffmm::core::devices::Virtual_Device* m)
+static void showMonitoredDevices()
+void CheckStatus(in ffmm::core::devices::Virtual_Device* m)
+<<pointcut>><<advice>> +device_constration()
+<<pointcut>><<advice>> +device_destrution()

Virtual_Device

+Virtual_Device(in :string, in :string, in :string)
+~Virtual_Device()
+SetCommunication_Bus(in :int) : int
+get_id() : int

Concrete aspects:
Pointcuts to capture faults
by intercepting change to
device status.

<<aspect>>VirtualDevice1_FaultDector

-faultTable:VirtualDevice1_FaultTable
-decoder:VirtualDevice1_FaultDecoder

+<<pointcut>>+checkOperationalStatus()
+<<pointcut>>+checkForValidConfiguration()

<<aspect>>VirtualDevice2_FaultDetector

-faultTable:VirtualDevice2_FaultTable
-decoder:VirtualDevice2_FaultDecoder

+<<pointcut>>+checkOperationalStatus()
+<<pointcut>>+checkForValidConfiguration()

VirtualDevice1

+createDevice()
+deleteDevice()
+set_Communication_Bus(in : int) : int
+startDevice()
+stopDevice()
+resetDevice()

VirtualDevice2

+createDevice()
+deleteDevice()
+set_Communication_Bus(in : int) : int
+startDevice()
+stopDevice()
+resetDevice()

Device1

Device2

Concrete virtual devices
acts as fault sources.
Each time a fault happens
the devices status
changes triggering the
aspect logic to be
executed.

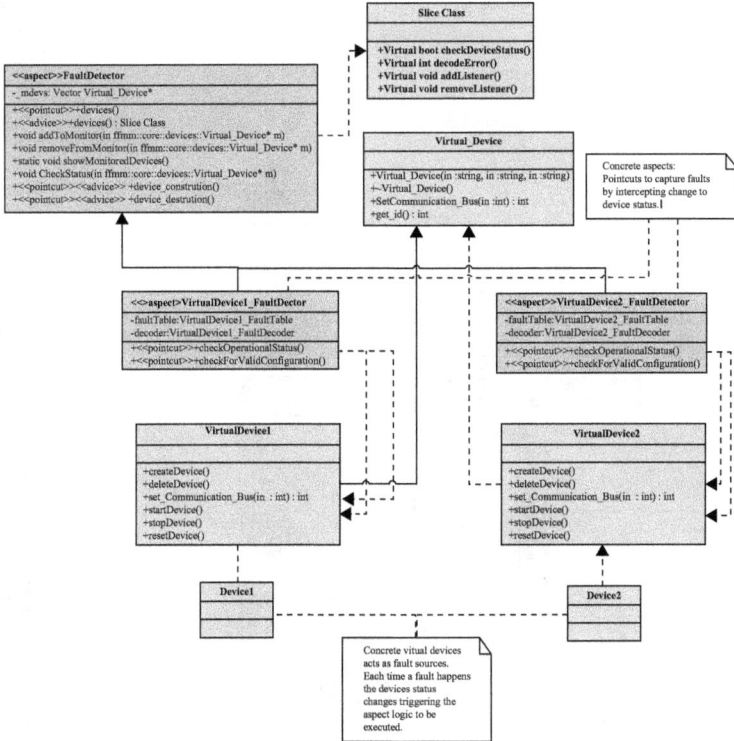

Figure 4.9. An excerpt of the hierarchy of the *Fault Detector*.

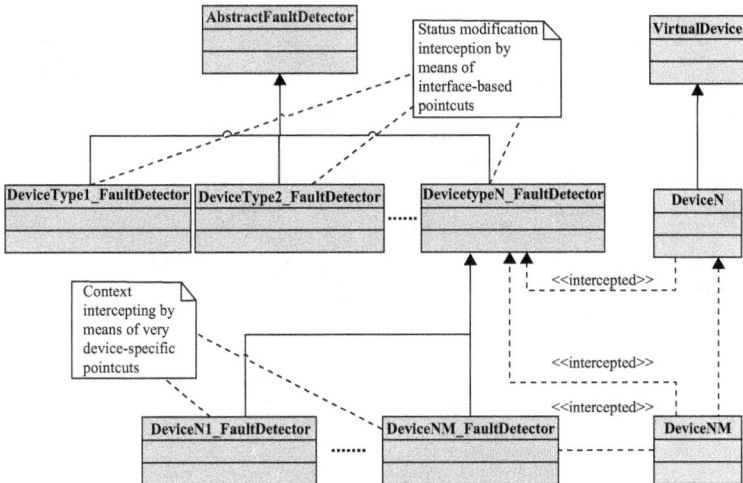

AbstractFaultDetector

Status modification
interception by
means of
interface-based
pointcuts

VirtualDevice

DeviceType1_FaultDetector **DeviceType2_FaultDetector** ...... **DevicetypeN_FaultDetector** **DeviceN**

Context
intercepting by
means of very
device-specific
pointcuts

<<intercepted>>

<<intercepted>>

<<intercepted>>

DeviceN1_FaultDetector ...... **DeviceNM_FaultDetector** **DeviceNM**

Figure 4.10. Levels of faults interception.

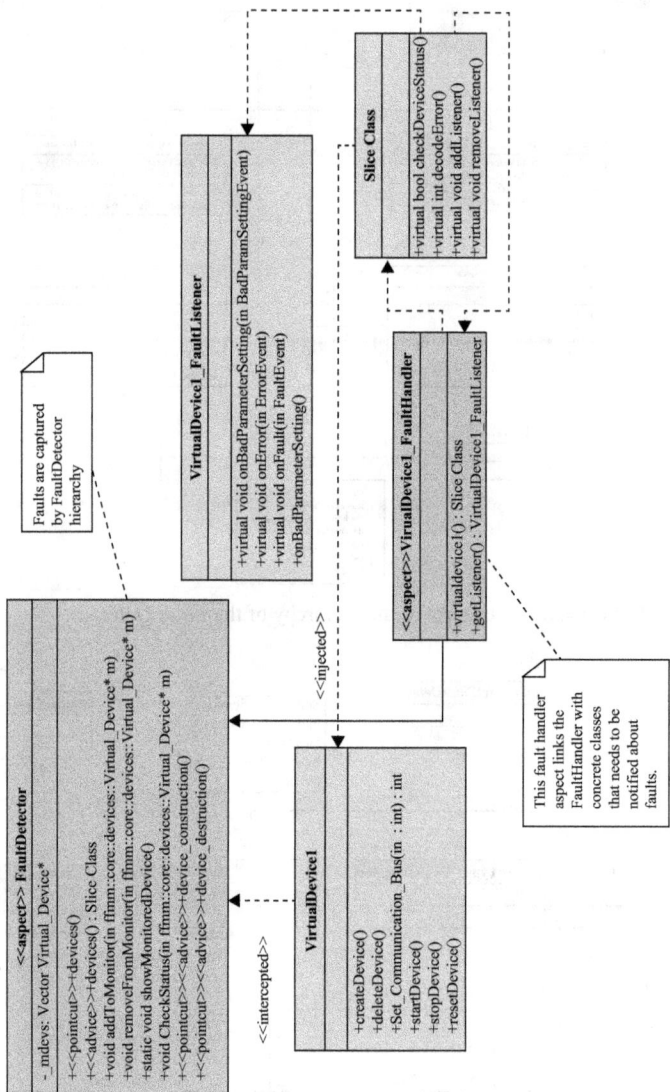

Figure 4.11. Fault notification publish-subscribe architecture.

The top level includes abstract aspects implementing the fault detection logic reusable in concrete subaspects. In Figure 4.11, the aspect-mapping layer of the fault notification is shown. The services necessary to associate dynamically the handlers to fault sources in the measurement system are provided. The subaspects of the *FaultHandler* aspect have the responsibility for making aware the concrete classes (like the *TestManager*, responsible of performing the test session) of the faults that happens in the system.

This solution allows fault handling logic to be reused in the super-aspects and does not force concrete classes in the system to implement fault handling code. Any component in the system can react to specific faults that occur anywhere in the system and perform the needed actions to handle them. Moreover, since concrete classes (*TestManager* or any other components interested in monitoring faults) are oblivious of being faults handlers, the monitoring relationships can be changed by simply acting on aspect mapping [10]. Commonalities among different fault handling logics can be factored out in the aspects, while multiple observations of different kinds of faults can be easily accomplished by defining several mapping aspects for a single concrete class.

4.4 SYNCHRONIZER

In this section, a PN-based approach to software synchronization in automatic measurement systems is presented. Tasks are synchronized by means of a PN modeling an execution graph, where nodes represent tasks, while arrows among nodes point out time succession among the corresponding tasks [11]. This allows software synchronization to be abstracted above the code level, by leaving the test engineer to work at a more intuitive level.

In particular, in the following, (a) a short introduction to *software synchronization in measurement automation*, (b) the *basic ideas* underlying the component design, and (c) the *design* with an *evolution example* of the Synchronizer are described.

4.4.1 SOFTWARE SYNCHRONIZATION IN MEASUREMENT AUTOMATION

In automatic measurement systems, usually asynchronous tasks have to be run concurrently on the same platform. A crucial issue is the capability of assuring a proper synchronization to the measurement procedure. Whereas severe time constraints require a dedicated hardware, at software level, task's interaction often requires programming strategies capable of dealing with events generated asynchronously and notified to the processes once a synchronization point is reached [12, 13]. Today, software synchronization is a technique used widely, and emerging application areas for cost-effective dependable systems will further increase its importance [14]. Typical examples of software synchronization are: (a) one or more tasks must wait for the termination of other tasks before starting, (b) events have to be notified to one or more tasks, and (c) a task has to be enabled to start when a particular event is triggered, and so on.

In the past, various types of "synchronization objects" have been used in coordinating the execution of multiple threads and processes. A common type of synchronization object is a *mutex* (short for *mutual exclusion*) [11, 15]. A mutex may be used to guarantee exclusive access to a shared resource, typically by controlling access to the resource through operations of "lock" and "unlock." This technique of waiting for a mutex is often called "blocking" on a mutex, because the thread or process is actually blocked and cannot continue until the mutex is released. Other types of synchronization objects include *semaphores* and *queues* [15, 16].

Generally, test engineers managing automatic measurement systems are not skilled programmers. Thus, they often find it difficult to implement the execution of software synchronization properly by using objects such as mutex, semaphores, or rendezvous [17]. Therefore, any systems and methods for supporting the synchronization of measurement tasks turn out to be very useful. In particular, it would be desirable to abstract synchronization above the code level, so that the test engineer can work at a more intuitive level [11].

Recently, a new generation of frameworks supporting software production for test applications is arising [18]. In particular, at a commercial level, with TestStand of National Instruments [19], steps, such as individual tests, measurements, actions, or commands can be automated in a sequence, but not in parallel or in event-driven configuration. At research level, in the proposal of the consortium Tango [20], if the test engineer wants to decompose its application in multiple tasks, he will be forced to design a client application adequately, by managing threads, semaphores, and so on. In the Extensible Measurement System (EMS) of FermiLab [21], the application description language (ADL), a proprietary dialect of XML, allows sequences of control events to be described. However, only common actions, that is, initialize or start, can be executed in parallel.

Recently, PNs, graphical and mathematical modeling tools applicable in different environments and in measurement systems also, have been a focus of scientific interest in (a) evaluating CAN-bus performance [22],

(b) monitoring systems based on microcontrollers [23], (c) failure monitoring systems for protection in distribution network [24, 25], (d) modeling and analyzing test systems [26], (e) detecting and diagnosing faults in industrial environments [27], and (f) some measuring medical applications [28]. PN algorithms have been also used successfully in the design of distributed measurement systems [29], or, more specifically, in modeling its data acquisition modules [30]. In these applications, they permit to describe and model information processing systems characterized as concurrent, asynchronous, distributed, parallel, nondeterministic, and stochastic [31]. Each part of the measurement system is easily modeled at a high level, leading to a whole library of partial models considering time dependencies within the system.

4.4.2 BASIC IDEAS

A measurement procedure includes actions to be performed sequentially or concurrently. Let's assume the test engineer responsible for writing the measurement script corresponding to the procedure (Figure 2.1) does not have software skills. Suitable tools for scheduling the execution of measuring tasks in a simple and intuitive way have to be provided. The *Synchronizer* helps the test engineer to think at a high level, in terms of: (a) "the task A has to be executed first," (b) "task B has to be executed after task C," and (c) "the task D has to be executed when the event E occurs."

On this basis, the main leading concept of the *Synchronizer* design is to make available to the test engineer a software component for scheduling the execution of a procedure at a high level, by modeling sequential and parallel executions of tasks, tracing their dynamic status, and determining the available task, step by step. In this way, the test engineer can (a) subdivide a generic measurement application in different measurement tasks and (b) determine their order of execution, without worrying about details of time synchronization.

With this aim, the *Synchronizer* is based on a Petri net, allowing the dynamics of execution to be organized step by step fully transparently to the test engineer. In literature, Petri nets have been used either for assessing performance of systems or for carrying out simulations. In particular, several years of research have established Petri nets as a powerful modeling formalism. Their formal semantics make them suitable for complex concurrent processes' description, for software performance evaluation [32, 33], for system simulation [34], for project modeling and simulation [35], in the field of communication networks [36], and other several fields.

In conceiving the *Synchronizer*, their use turns out to be useful at the exploitation level:

1. Preliminarily, in a static way, in order to store the execution graph, defined by the test engineer in the measurement script when each declared task is executed.
2. Successively, in a dynamic way, the active properties of Petri net's are exploited for tracing the tasks already executed and, by leaving the net to evolve, to obtain the list of tasks ready for execution.

The major aim of the framework is to make software production easier. The *Synchronizer* simplifies a step-wise decomposition of a measurement application by allowing measurement task-level details to be separated from high-level overviews. This makes the measurement procedure specification easier by using the divide-and-rule principle.

Therefore in synthesis, the main basic idea of the *Synchronizer* is just the twofold uses of the Petri nets: At application level, for generating measurement applications in a stand-alone general-purposes module for task synchronization, and, at exploitation level, in combining static and dynamic properties for separating static easy task description from complex concurrent management.

In particular, the concept of Petri nets-based *Execution Graph* is utilized: (a) a node represents a task or an event, (b) an arrow from a task node A to a task node B implies that the task node B has to be executed after the task node A is completed, and (c) an arrow from an event node E to a task node C implies that the task node C has to be executed when event E occur [11].

Two key software components, the *Test Manager* and the *Synchronizer*, are conceived. In particular, the *Test Manager* is responsible for

- requiring the list of tasks to be executed by the *Synchronizer*;
- starting the execution of each *Measurement Task*, by notifying the *Synchronizer* of this;
- detecting when a measurement task ends its execution, by notifying the *Synchronizer* also of this.

The *Synchronizer* [11] is responsible for

- managing a data structure implementing the *Execution Graph*;
- getting the notification of a task start and termination and evolving the *Execution Graph* status consequently.

From a dynamic point of view, at each execution step:

1. The *Test Manager* asks the *Synchronizer* for the list of executable tasks.
2. The *Synchronizer* checks the status of the *Execution Graph*, and provides the *Test Manager* with the list of executable tasks.
3. If the list is empty and no other tasks are in execution, the measurement application is terminated.
4. If the list is empty, but other tasks are in execution, the procedure skips to step 7.
5. If the list isn't empty, the *Test Manager* launches the execution of each task in the list, and notifies the *Synchronizer* of each execution.
6. When the *Synchronizer* receives a notification from the *Test Manager*, it evolves the status of the *Execution Graph*.
7. The *Test Manager* waits for the end of a task; then, it notifies the *Synchronizer* of this event.

When an event occurs, the *Synchronizer* evolves the status of the *Execution Graph*.

In Figure 4.12, a straightforward example, highlighting the working mechanism of the conceived *Execution Graph*, is shown:

1. The measurement begins with the execution of the task T_0.
2. When T_0 is completed, the task T_1 is executed.
3. When T_1 is completed, the tasks T_2 and T_3 are started simultaneously.
4. When the event E_1 is triggered, for example, during the execution of T_1, the tasks T_4 and T_5 are started simultaneously, and
5. So on.

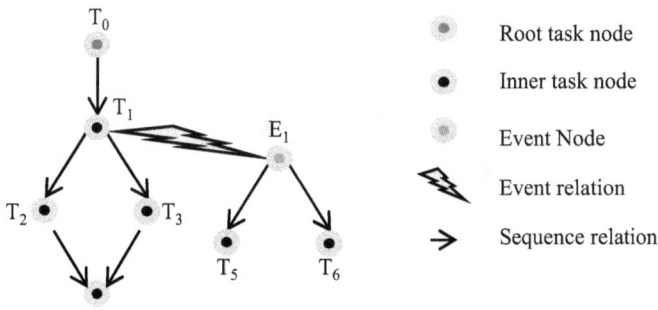

Figure 4.12. Working example of the *Execution Graph* [11].

In the measurement script, the *Execution Graph* is codified by the test engineer through the following commands provided by the *Test Manager*:

- ADD_TASK(task name)
- ADD_TASK_AFTER_TASK(previous task name, following task name)
- ADD_TASK_AFTER_ EVENT(event name, task name)

With respect to the example pointed out in Figure 4.12, the test engineer, after the definition of the tasks separately, defines the *Execution Graph* by means of a set of code lines as shown in Figure 4.13.

Another leading idea of the *Synchronizer* is to model the *Execution Graph* by means of a Petri net aimed at presenting simultaneously control and data flows in a concurrent system [17]. Its graphical representation (Figure 4.14) is a dual graph containing two types of nodes,

```
ADD_TASK(T_0)

ADD_TASK_AFTER_TASK(T_1, T_0)

ADD_TASK_AFTER_TASK(T_2, T_1)

ADD_TASK_AFTER_TASK(T_3, T_1)

ADD_TASK_AFTER_TASK(T_4, T_2)

ADD_TASK_AFTER_TASK(T_4, T_3)

ADD_TASK_AFTER_EVENT(T_4, T_3)
```

Figure 4.13. Code lines for the Execution Graph definition [11].

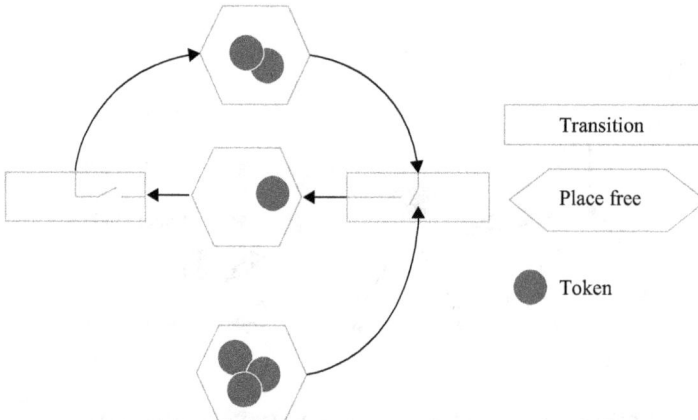

Figure 4.14. Example of a Petri net [11].

called "places" and "transitions." Only nodes of different types can be connected by directed arcs. The places (symbolized as circles or ellipses) represent states, while the transitions (rectangles) simulate events. The places in the network contain tokens, represented by dots. Displacement and flow of the tokens determines dynamics in the system, that is, its changes in time.

The PN of the *Synchronizer* is extended by a Labeled Petri net (LPN) [37] in order to offer a more consistent way of the measurement procedure description, as well as to simplify its modeling and analysis. In the LPN, each place and transition has an associated label in order to allow different classes of places and transitions to be modeled and managed. In synthesis, the LPN allows

- a task state (e.g., in execution, terminated, ...);
- a temporal relation between the execution of two tasks (e.g., run task *T2* after task *T1*); and
- a relation between a task execution and events (e.g., run task *T5* after event *E1*), to be modeled easily and consistently.

4.4.3 DESIGN

On the basis of the aforementioned basic ideas, the design of the *Synchro-nizer* is aimed at satisfying the following requirements:

- Building the execution graph, by adding a node, an event, or an arrow.
- Querying the execution graph, by determining the executable nodes, the end-node, and the loop detection.
- Updating the execution graph, by forcing the execution graph dynamics: Execute a node, terminate a node, freeze a node, unfreeze a node, notify an event, and set an executable node.

The previous requirements are satisfied by means of the following classes (Figure 4.15):

- *PetriNet*, supplying all basic methods to manage a Petri net.
- *Place*, allowing tokens, labels, inner, and outer arcs to be managed.
- *Transitions*, allowing transitions to be enabled or disabled, as well as labels and inner and outer arcs to be managed.
- *Synchronizer*, supplying all methods to manage the *Execution Graph*.

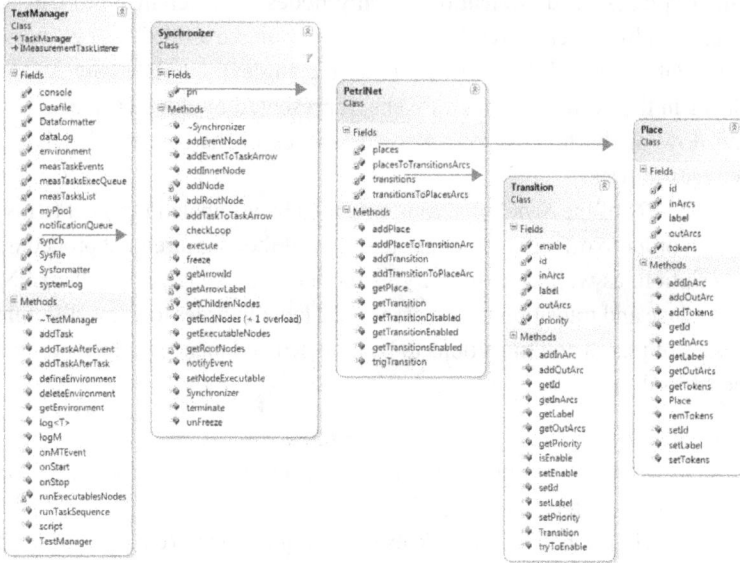

Figure 4.15. Architecture of *Synchronizer*'s classes [11].

- *TestManager*, supplying all methods to manage the execution of measurement tasks [11].

Basically, the class *PetriNet* provides all the methods necessary to build a generic PN (*addPlace*, *addTransition*, and so on), by using the basic classes *Place* and *Transition*.

The class *Synchronizer* provides all the methods to build and to manage the *Execution Graph* by using a private Petri-net object. As an example, the method *addRootNode* permits to add a node (task) to be executed as the first, while the method *getRootNodes* permits to obtain all the root nodes. In this way, the *Synchronizer* hides the details of the PN and performs high-level methods for using the *Execution Graph*. In particular, three kinds of nodes are provided: (a) the event node, representing a task to be executed when a particular event occurs; (b) the task node, representing a task to be executed after another node (task) is terminated; and (c) the root node, a special task node to be executed when the measurement application starts.

In Figure 4.16, details about the implementation of the event and task nodes are shown:

- The **event nodes** are characterized by a transition named "trig" connected by an inner arrow to a place named "triggered." One or

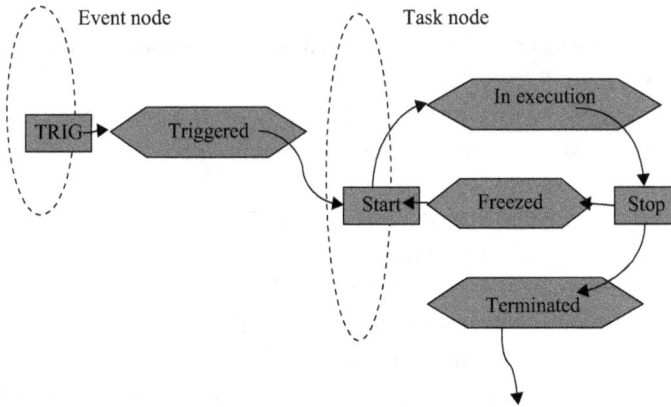

Figure 4.16. Implementation of the Execution Graph entities (nodes and arrows) [11].

more outgoing arrows allow the *EventNode* to be connected to one or more *TaskNodes*.

- The **task nodes** trace three different states by means of three places, named "in execution," "freezed," and "terminated," and two transitions, named "start" and "stop." Inside the *TaskNode*, the arrows model the right sequence of the task states and one or more outgoing arrows allow the *TaskNodes* to be connected to other nodes.

- The **root nodes** are particular task nodes, characterized by the absence of incoming arrows. In fact, the task associated with the root nodes starts at the beginning of the measurement application.

4.4.3.1 Evolution Example

In this section, a straightforward example, highlighting how a generic *Task Manager* can use the *Synchronizer*, is described. Furthermore, the LPN dynamic evolution for tracing the execution status of each task is illustrated, by highlighting specifically how, on demand, the list of the tasks available for the execution is provided.

Let the sequential and parallel executions of tasks T_1, T_2, T_3, and T_4 be modeled by using the *Execution Graph* [11]. The following actions are carried out (Figure 4.17a):

1. When the *Task Manager* is ready to carry out the *Execution Graph* (runTasks), as a first step, the list of the nodes to be executed (getExecutableNodes) are required to the *Synchronizer*, at this time, they are the root nodes (only T_1 in Figure 4.17b).

2. The *Synchronizer*
   - checks on its LPN if there is a "start" transition enabled;
   - finds T_I "start" transition enabled;
   - returns a list with the task T_I to the *Task Manager* (nodeList[T_I]).
3. The task manager notifies the *Synchronizer* of executing T_I, (execute (nodeList[T_I])).
4. The *Synchronizer* modifies (Figure 4.17c) the execution status of T_I (from "ready" to "in execution") by
   - disabling the "start" transition of T_I;
   - adding a token to the "exec" state of T_I;
   - enabling the "start" transition of T_I.
5. When the *Task Manager* catches the termination event of T_I (Figure 4.18a), it also notifies the *Synchronizer* of the termination of T_I.
6. The *Synchronizer*, modifies (Figure 4.18c) again the execution status of T_I (from "in execution" to "terminated") by
   - removing the token from the "exec" state of T_I;
   - disabling the "stop" transition of T_I;
   - adding a token to the "wait" state of T_I.

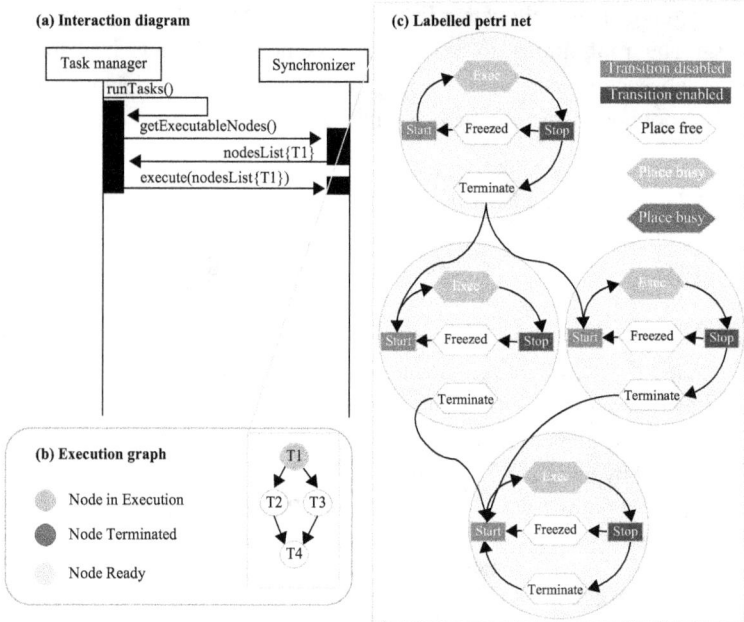

Figure 4.17. A generic *Task Manager* uses the *Synchronizer* to select an executable task [11].

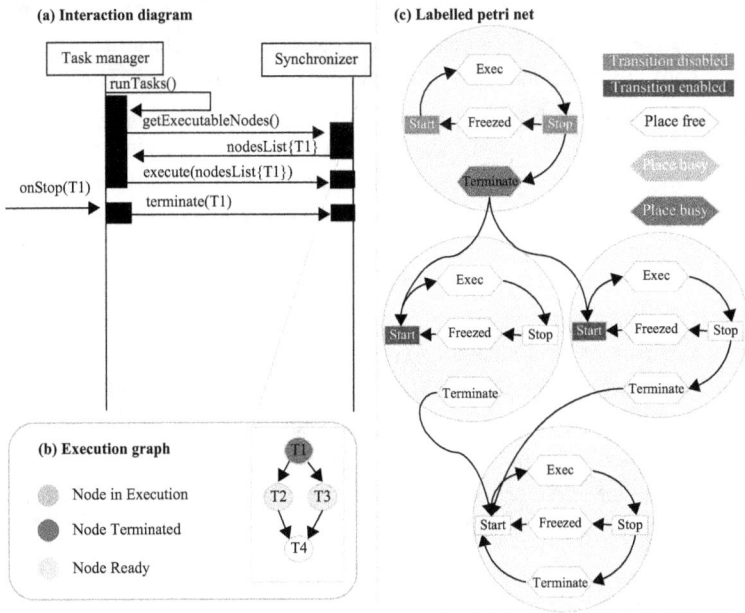

Figure 4.18. A generic task manager uses the *Synchronizer* to trace the change of the tasks execution status [11].

7. Moreover, the *Synchronizer* modifies to "ready" the execution status of all tasks whose execution starts after the termination of T1, that is tasks T_2 and T_3 by
   - enabling the "start" transition of T_2;
   - enabling the "start" transition of T_3; and
8. So on, like in the step 1, for T_2 and T_3 [11].

4.5 MEASUREMENT-DOMAIN SPECIFIC LANGUAGE

In Chapter 2, domain specific languages (DSLs) are defined as languages specialized for a given domain. While general-purpose languages are designed to describe whatever procedures, DSLs are limited: One will never write an application with 3D graphics in SQL, for example. Conversely, within their own domain, DSLs performs better than general-purpose languages. DSL abstractions shorten the development cycle and makes maintenance easier. In some cases, DSLs can be specialized enough to allow them to be used even by nonprogrammer domain experts.

In this section, a measurement-domain specific language, based on a model-driven paradigm for measurement test procedure definition, instrument configuration, and tasks synchronization, is presented. This formal language, particular for a specific measurement field (e.g., magnetic measurements, automotive tests, and so on), aims at specifying complete, easy-to-understand, -reuse, and -maintain applications efficiently and quickly by means of a script. The script is checked and integrated into the existing software framework automatically by a specific parser-builder chain, in order to produce the measurement application. Constructs for abstracting key concepts of the domain allow the test engineer to write more concise and higher-level programs by natural language-like sentences in a shorter time without being a skilled programmer [38].

In particular, in the following, (a) a short *introduction* to languages in measurement automation, (b) the *basic ideas* underlying the design, and (c) the *architecture* by detailing its *Parser* and *Builder* of the measurement-domain specific language are described.

4.5.1 INTRODUCTION

In the context of the research trend on programming languages for developing automatic measurement and test systems, graphical environments characterized by graphical icons and wires symbolizing the data flow (such as LabVIEW by National Instruments [39]) are very popular. The underlying approach of the programming language (e.g., the G for LabVIEW [40]) emphasizes the objects involved in the application and the data exchange, with less emphasis on the representation of the temporal sequence, even if both serial and parallel executions are possible. Conversely, in scientific and academic communities, scripting languages are commonly used and preferred [41]. They point out the order of operations and allow the temporal constraint in the measurement application to be managed easily. Advanced software techniques (e.g., software framework, scripting, new programming languages) have been exploited recently in the area of magnetic-field measurements [21]. These solutions lead to maximum software-component reuse, but final users have to manipulate code written in extensible markup language and Python. Successful examples of new programming languages for improving measurement applications and reducing the development effort can also be found in network performance testing [42] and in monitor and control systems [43].

In this section, these points are taken into consideration and model-driven-development (MDE) concepts are exploited [44] for highlighting a measurement-domain-specific scripting language (MDSL) aimed at defining test procedure scripts, configuring instruments, and synchronizing measurement tasks [45].

4.5.2 BASIC IDEAS

The MDSL goal is achieved by fulfilling the following requirements of providing instructions for:

1. defining logical, numeric, and temporal conditions;
2. defining conditional branching;
3. defining events based on measurement values or attribute changes, time changes, external event notifications, or user inputs;
4. subscribing and unsubscribing to events and responding to them with behaviors including messages to users or commands;
5. setting, configuring, and commanding the devices;
6. enabling, configuring, and disabling software services (e.g., log, fault detection, etc.); and
7. interacting with the user through graphical or textual interfaces [45].

The basic idea for conceiving a DSL for measurement applications arises logically from the analysis of a typical measurement procedure [45]. Usually, a measurement description should be expressed by a few specialized commands concerning device management, the temporal sequence of actions, and data treatment. At the abstraction level of the commands for physical (i.e., hardware devices connected by a bus) or virtual objects (purely software components providing some services), the measurement description can be seen as a script. The script is a relatively short program composed of a sequence of synthetic instructions controlling the operation and performing a work task all in one batch. The power of a script organization of the test program, instead of a graphical representation (such as in LabVIEW), is mainly related to the immediacy of translating a well-defined procedure in a logically ordered synthetic set of high-level commands. The approach presented here consists of defining a new DSL for measurement application, the MDSL, for producing batch applications within a software framework. As with any programming language, MDSL has its own lexicon, syntax, and semantics. A low level of complexity is provided by keeping the lexicon bounded and often using mnemonic words from the domain, and by simplifying the syntax through straight-

Figure 4.19. MDSL process according to the pattern "source-to-source" [38].

forward command structures, for example, the commands for piloting a device are divided into categories of similar meaning (e.g., settings, execution, and so on) and with the same use.

The new language is combined with the software basis of the existing framework, by decoding the MDSL code through a custom parser (i.e., a measurement-domain-specific parser) first, and then by translating it to the target language (C++) through a builder (i.e., a measurement-domain-specific builder). Such an approach can be ascribed to the creational pattern "source-to-source transformation" [46] (Figure 4.19).

4.5.3 ARCHITECTURE

The MDSL is based on a separated semantic model [43], conceived as a part of a framework domain model. Generally, the semantic model consists of a network of concepts and the relationships among those concepts. In this context, concepts are particular ideas or topics from the user perspective with which the user is concerned. In the MDSL, the semantic model describes the measurement-test-procedure core structure and determines its behavior. The model is separated from the MDSL in order to:

1. think about the semantics of the domain without getting tangled up in the MDSL syntax or parser;
2. be able to test the semantic model by creating objects in the model and manipulating them directly;
3. follow an incremental approach by starting with a simple internal MDSL and then adding an external MDSL; and
4. be able to evolve the model and the language separately: A change in the model can be explored without changing the MDSL by adding the necessary constructs, or new syntaxes for the MDSL can be experimented by just verifying the creation of the same objects in the model.

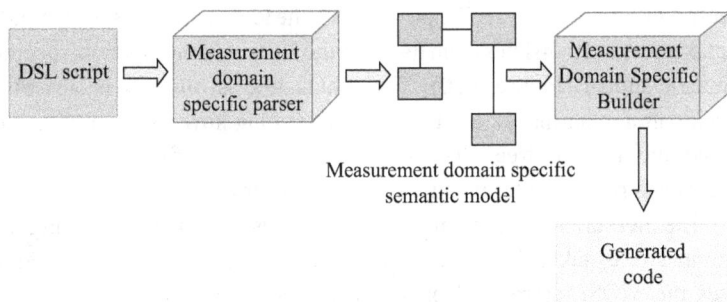

Figure 4.20. Semantic model and MDSL architecture [38].

The separation of the semantic model and the MDSL syntax mirrors the separation of the domain model and presentation suggested in [47]: The MDSL can be thought of as another form of user interface.

In Figure 4.20, another benefit of using a semantic model is highlighted: The code generator or builder is decoupled from the parser, and a code generator can be written without having to understand anything about the parsing process. This makes possible independent testing.

4.5.4 MEASUREMENT-DOMAIN SPECIFIC LANGUAGE PARSER

Parsing is a strong hierarchical operation [48]. When a text is parsed, the chunks are arranged into a tree structure. Let us consider the simple structure of a list of two events in a measurement procedure where an encoder generates a trigger signal for a motor controller in order to start the rotation of a shaft:

```
Events
    EncoderBoard_StartTrigger EB_ST
    MotorController_StartRotation MC_ST
End
```

In this composite structure, a list contains events with their own name (e.g., EncoderBoard_StartTrigger) and code (EB_ST). The explicit notion of an overall list is missing, and each event is still a hierarchy of events with their own name symbol and code string.

Therefore, the MDSL script is represented as a hierarchy, that is, the "syntax tree" (or "parse tree"). A syntax tree is a much more useful representation of the MDSL script than the words [38]: It can be manipulated in many ways by walking up and down the tree. Basically, the measurement-domain-specific parser reads the textual MDSL script, builds syntax trees, and translates them to obtain the measurement-domain-specific semantic model (Figure 4.20). The syntax tree is built by means of a specific grammar, that is, a set of rules describing how a stream of text is turned into a syntax tree. Grammar consists of a list of production rules, where each rule has a term and a statement of how it gets broken down.

The measurement-domain-specific parser is designed by defining the grammar rules taken from the measurement domain. Such a parser analyzes the MDSL script and provides the measurement-domain-specific builder by an exhaustive specific semantic model.

4.5.5 MEASUREMENT DOMAIN SPECIFIC LANGUAGE BUILDER

The model-driven architecture (MDA) is an approach to the software development of the Object Management Group [49]. A tool implementing the MDA concept allows developers to produce models of the application and generate the corresponding code for a target platform by means of suitable transformations. The MDA is a sound groundwork for automatic code generation. As a matter of fact, the basis of automatic code generation is to read project artifacts (such as class diagrams, activity diagrams, and requirement documents) and turn them into a meaningful and correct source code. The implementation of automatic code generators relies on the fact that most artifacts are created in the early stages of software. These artifacts are repetitive and have design patterns; thus, they can be automated. Most simple implementations of automatic code generators use only the class diagram to create a source code. Class diagrams have been the easiest to implement because of the inherited design pattern to Object-Oriented languages such as Java and C++. In general, code-generation techniques give rise to the production of a nonoptimized code. However, in a delimited command vocabulary such as the measurement domain, a specialized builder allows an easiness and efficiency tradeoff, not too costly in terms of computation.

In the MDSL architecture (Figure 4.20), the measurement domain-specific builder is responsible for the code generation according to the MDA approach. From the domain-semantic model instance produced by the measurement-domain-specific parser, specific situations are represented in order to create the corresponding code. The builder, rather than a class diagram, takes from the parser the structured semantic model as input. It is able to recognize the classes, the methods, and the sequence of actions to be carried out, as well as to translate the model into the final code.

4.6 ADVANCED GENERATOR OF USER INTERFACES

In this section, a model-based approach, the Model-View-Interactor Paradigm [50], for automatically generating user interfaces [51] in a software framework for measurement systems is illustrated. The paradigm is dedicated to the "interaction" typical in a software framework for measurement applications (Figure 2.1): The final user interacts with the automatic measurement system executing a suitable high-level script previously written by a test engineer. According to the main design goal of frameworks [52], this approach allows the user interfaces to be separated easily from the application logic for enhancing the flexibility and reusability of the software.

In particular, in the following, (a) a short introduction to *graphical interfaces in measurement automation*, (b) the *Model-Viewer-Interactor paradigm*, and (iii) the *Graphical User Interfaces engine* are described.

4.6.1 USER INTERFACES IN MEASUREMENT AUTOMATION

In the first phase of designing a measurement software framework, the test engineer prepares a script where the required test procedure is expressed formally and synthetically by the MDSL. Then, the framework processes the script suitably and generates an executable measurement software application. In the second phase [52], the final user (a) executes the software application, (b) interacts with the software by providing at runtime the required input, and (c) finally starts the measurement process on the devices. The application user needs to interact with the software application through a convenient graphic interface to carry out the measurement procedure easily and quickly. Therefore, the test engineer should deal with its implementation.

Such as, for most interactive applications, producing an attractive Graphical User Interface (GUI) for a measurement software framework is not a trivial task [51, 52] because graphical representations depending on run-time data cannot be drawn in advance. The powerful GUI libraries offered by operating systems can be used of course, but the offered level of abstraction is, in general, rather low. Therefore, a visual editor, such as many commercial programming environments, should be used. Such tools turn out to be very user-friendly, although integrating the provided and existing codes is in general not easy.

Summarizing, a visual editor is a useful tool for simple applications, but for more complicated professional measurement applications, the test engineer has to still struggle with a low-level programming code devoted to GUIs. In addition, the quality of self-made GUI development depends strongly on the experience of the designers as well as on their skills in the platform and development graphic tools.

For these reasons, user interface generation has been object of research for many years, sometimes under the diction of "model-based user interfaces" [51]. As a matter of fact, interfaces are generated by dividing the application domain in models. The original contribution of this research field has been to allow programmers, as test engineers, not typically trained to design interfaces, to produce user interfaces customized to their own applications.

On the other hand, the main feature of automatic techniques for generating interfaces is to allow the designer to specify them at a very-high level, with the details of the implementation provided by the system [51]. Nevertheless, this approach is very unspecific and further effort is required to tailor the model to a definite context, such as the frameworks for measurement software products. Designing interaction rather than interfaces attempts to enhance the quality of the interaction between user and computer, fitting the key paradigm: "User interfaces are the means, not the end" [52].

In the context of measurement software products, LabVIEW [39] is very popular. It is a graphical programming environment used to develop measurement and test systems by using graphical icons and wires well symbolizing the data flow [40]. The input-output of the measurement program (called in such a context, the Virtual Instrument) is provided directly in the visual form of a GUI looking like the panel of an instrument. However, although powerful and intuitive from the GUI viewpoint, the approach of the programming language G exploited by LabVIEW is to not point out immediately the temporal sequence of the actions to be executed. Conversely, in a scientific and academic community, scripting languages are usually preferred [41].

In the following, an evolution of model-based user interface generation in a measurement software framework based on domain specific language for scripting [38], the Model-View-Interactor paradigm, is presented. This paradigm is mainly aimed at shifting the focus of the test engineer's work from the burden of implementing the interface to the conception of its abstract interactions.

4.6.2 THE MODEL-VIEWER-INTERACTOR PARADIGM

In the model-based approach to GUI generation [50], analysis, design, and implementation are based on a common repository of models. In this context, a model is a declarative specification of some single coherent aspects of a user interface, such as the appearance, the interaction with the user, and the layer of the underlying measurement application. The most important property of the model in this context is that it can be expressed in a highly specialized notation by focusing the attention on a single aspect of an interactive system [51]. This property makes the development and maintenance easier. Unlike conventional software engineering where designers construct artifacts whose meaning and relevance can diverge from the delivered code, in the model-based approach, designers first build models of critical system attributes, and then analyze, refine, and synthesize these models into running systems. This separation greatly simplifies and improves the quality of the development.

Early examples of model-based tools included Cousin [53] and HP/Apollo's Open-Dialogue [54], which provided the designer with a declarative language for listing the input and output requirements of the user interface. The system then generated dialogs to display and request data. These evolved into model-based systems, such as Mike [55], Jade [56], UIDE [57], ITS [58], and Humanoid [59]. They exploit specific techniques, such as heuristic rules, to automatically select interactive components, layouts, and other details of the interface.

Generating interfaces automatically is a very difficult task, because the corresponding automatic and model-based systems constrain significantly the kinds of interfaces they can produce. A related and very-common problem is that the generated user interfaces are generally not as good as those created by conventional and custom programming techniques. However, in specific domains with very few particular graphical requirements, such as the measurement domain, automatic techniques can be used successfully.

Model-based GUI generation relies on the principle that development and support environments may be built around declarative models[1] of a system [60]. Developers using this paradigm build the interface by

---

[1]*Declarative programming* is a software design approach focusing more on the computation logic rather than its control flow [62]. Main effort is to define what the code should do in terms of the problem domain, rather than as a succession of programming instructions. This is in contrast with *imperative programming*, where algorithms are implemented by categorical steps. Examples include database query languages (e.g., SQL, XQuery). The program of a *declarative model* includes directly equations, rather than imperative statements, defining ("*declaring*") behavioral relationships. In this case, the procedure carries out algebraic operations to best express the solution algorithm. Typical examples are Modelica and Simile.

specifying declarative models, rather than writing a program. Generally, for any interactive system, three kinds of models can be derived [61]:

- Presentation models, specifying the appearance of user interfaces in terms of their widgets and the related behavior.
- Application models, specifying the parts (functions and data) of applications accessible from the user interface.
- Dialogue models, specifying end-user interactions, their order, and how they affect the presentation and application.

In the following, (a) the *basic concepts*, (b) the *View*, (c) the *Interactor*, and (d) the *Model* of the Model-View-Interactor paradigm are illustrated.

4.6.2.1 Basic Concepts

The approach presented here to generate interfaces in measurement system frameworks starts from a fundamental consideration: Usually test engineers are not trained to design interfaces, but at the same time they would like to maintain a high level of usability in measurement applications. In this case, test engineers are responsible mainly for preparing test scripts [6], where the interaction between measurement application and final user are described at high level [38], without any indication of GUI aspects.

Therefore, the main concept underlying the approach presented here is the interaction between the user and the GUI. Interaction is a kind of action occurring when two or more objects have an effect upon one another. Examples of simple interactions in measurement software are reading a user input or displaying a value. Test engineers are prevented from dealing with raw graphical characteristic of software measurement system, by separating functional from look aspects of the interface. Accordingly, the architecture is organized by a three-way decomposition: (1) the parts representing the model of the underlying application domain, (2) the way the model is presented to the user, and (3) the way the user interacts with it. This is called the *Model-View-Interactor* approach [63] (Figure 4.21), derived as an evolution of the Model View Controller and Model View Presenter [64].

The *Model* represents the model domain, and, in case of measurement software framework, is constituted by the core classes. The *Views* consist in the aspect of the generated GUI, defined by a GUI expert, completely transparent to the test engineer using the framework. In particular, the

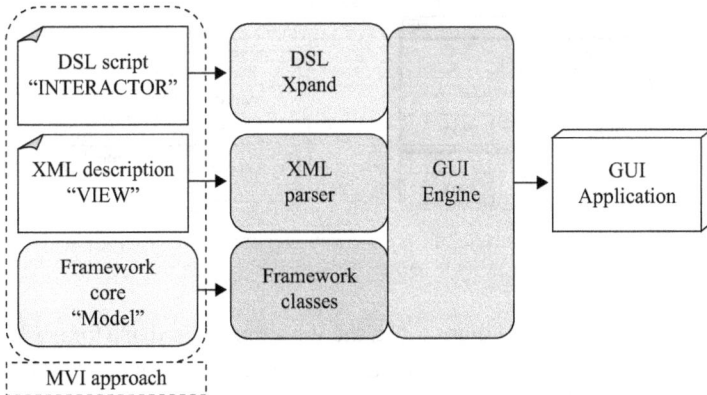

Figure 4.21. *Model-Viewer-Interactor* (MVI) approach.

GUI expert defines a set of presentation models used to generate the final user interface. The *Interactor* represents the tie between model and view, by making available a different component specifying the GUI desired behavior.

In this way, the test engineer can define the interaction between the measurement application and the user by means of a set of specific objects: The Graphic Interactor Component (Section 4.6.2.3).

4.6.2.2 View

The view description is an XML-file containing all the presentation features of the GUI. XML stands as a solution for the standardization of the interoperability between applications. Therefore, XML-based languages can be employed to define user interfaces. They are the XML-compliant user interface definition languages (XML-UIDL) [65] and their advantage is to be transparent to different interface technologies and to provide a homogeneous resource for heterogeneous ways of interaction.

Generally, at a graphical level, the user-interface content can be organized in areas, represented usually by a rectangular bounding shape. These rectangular areas are referred to as a box. Graphical user interface layouts can be seen as a container subdivided in to boxes, where the graphic components (text editor component, buttons, menu item, and so on) can be placed. One box can contain others boxes, and so on. Two types of boxes can be distinguished: (a) horizontal boxes (HBox), with elements aligned horizontally, and (b) vertical boxes (VBox), with elements aligned vertically.

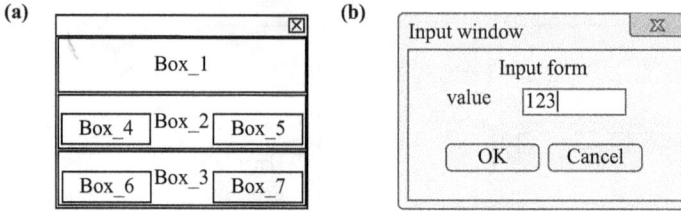

Figure 4.22. Example of (a) a View model and (b) its final aspect.

As an example, in Figure 4.22a, the View model used in a form asking for an input value to the final user is considered. The layout as a whole is formed by 3 VBox (Box_1, Box_2, and Box_3):

- Box_1 will contain a text component for the title.
- Box_2 is formed by two HBox: Box_4 and Box_5. The former will contain a text component for a description, and the latter will contain a text editor component to read an input value.
- Box_3 is formed by two HBox: Box_6 and Box_7. Both of them will contain a button component. The button in Box_6 will command an action to confirm inserted input value, while button in Box_7 will command an action to discard the operation.

End user will see a form appearing as shown in Figure 4.22b. The View model in Figure 4.22a is stored as a declarative model, containing also the information about the GUI component properties (character font, text component color and text editor component positions, background and foreground colors, as well as further information related to the GUI aspect).

Basically the approach presented here is based on a database of different View models, associated to one or more interactive components, such as shown in the next section.

The view description example for a simple window is depicted in Figure 4.23. The view description file is written by the application engineer during the software development phase in order to define the GUI presentation look. Then, at run time, this file is read by the XML-Parser, and the framework uses the information to generate the graphical elements.

4.6.2.3 Interactor

The main aim of the Model-View-Interactor paradigm is to permit the test engineer to develop complicated GUI applications by a minimal effort

```
<?xml version="1.0"?>
<window id="box example" title="Input Form"
        <vbox>
                <staticText id="description"/>
        </vbox>
        <vbox>
                <hbox>
                <staticText id="label"/>
                <ctrlText id="control"/>
                </hbox>

        </vbox>
        <vbox>
                <hbox>
                <button id="ok" label="OK"/>
                <button id="cancel" label="CANCEL"/>
                </hbox>
        </vbox>
</window>
```

Figure 4.23. View XML description example.

and moreover without graphical knowledge. This aim is achieved mainly through the Graphic Interactive Component (GIC), allowing any customizable graphical component to be derived. This component encapsulates all common aspects of graphical components [66, 67]. It is generated automatically for any type T, for example, int, float, double, or more complex data types.

Namely, the GICT, defined for the type T, can

- be used by a test engineer to display automatically any value of type T;
- be used by a test engineer to plot on screen an array of type T;
- be used by an application user to view and edit some value of type T;
- communicate any value change made by the user or by the program with any other depending component.

For any concrete type T, the compiler is able to derive automatically an instance function of this meta-description for the given type.

The test engineer, in the writing phase of the MDSL script [38] (Figure 4.24), defines the component contained in the GUI and their input and output data by using the GIC components. Then, after building the script by means of the *DSL-Xpand* component, the framework can generate the application with the desired GUI.

4.6.2.4 Model

The Model is composed of data structures and classes of the framework involved in the GUI generation. A typical example is offered by the device

```
BEGIN SCRIPT " DSL example"
        Def  FDI "FDI1" with(2);
        Def GIC "InputParam";
        Capture InputParam with (2,"Parameter request","FDI bus:");
        BEGIN MTASK "Task1(start_procedure)"
        ...
        END MTASK
        BEGIN MTASK "Task2(flux_measurement)"
        ...
        END MTASK
END SCRIPT
```

Figure 4.24. MDSL script example.

classes related to the configuration step of the measurement procedure. During this phase, a broad interaction with the user is required to set up the devices. The data needed for the configuration are structured in the class definition where the data variables are preset for type and number by the application developers.

4.6.3 THE GRAPHICAL USER INTERFACE ENGINE

The classes' architecture allowing the automatic user interface generation is named *GUIengine* and is shown in Figure 4.25.

The *GUIengine* is composed of several classes: (a) *GIC*, providing the *TestManager* the input and output features without graphical details; (b) *GenericWindow*, giving the interface for all the frames; (c) *InputWindow*, *OutputWindow*, and *PlotWindow*, the concrete windows; and (d) *LayoutManager*, responsible for instantiating concrete windows defining the graphical features parsing the view description file and computing the dimension and position parameters [68]. The View is kept clear-cut from the Interactor by implementing the *GUIengine* complying with the *abstract factory* pattern,[2] often employed to separate the details of GUI implementation from its general use.

As an example, if the test engineer needs to ask as an input, an integer value at runtime, he will use the capture method of a GIC object in the measurement script [6]:

- Def GIC "*InputParam*"
- Capture *InputParam* with (*param,1,* "*Input form,*" "*value*")

---

[2]The *abstract factory* pattern allows a group of individual factories with a common theme to be encapsulated without specifying their concrete classes.

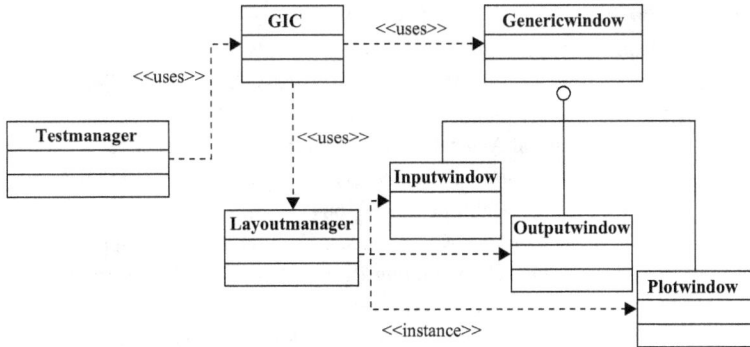

Figure 4.25. *Abstract factory* pattern for the GUI engine.

By inserting in the script only these instructions, a form is displayed, and the value entered by the user is stored in the variable pointed.

REFERENCES

[1] Riehle, D. 2000. Framework Design: A Role Modeling Approach (PhD Thesis, No. 13509), Zürich, Switzerland, ETH Zürich.

[2] Johnson, R.E., and B. Foote. June–July 1988. "Designing Reusable Classes." *Journal of Object-Oriented Programming* 1, no. 2, pp. 22–35.

[3] Riehle, D., and T. Gross. 1998. "Role Model Based Framework Design and Integration." In *Proceedings of the 1998 Conference on Object-Oriented Programming Systems, Languages, and Applications (OOPSLA '98),* 117–33. New York: ACM Press.

[4] Gamma, E., R. Helm, R. Johnson, and J. Vlissides. October 1994. *Design Patterns: Elements of Reusable Object-Oriented Software.* Boston, MA: Addison Wesley.

[5] Kiczales, G., J. Lamping, A. Mendhekar, C. Maeda, C.V. Lopes, J.M. Loingtier, and J. Irwin. 1997. "Aspect-Oriented Programming." In *Proceeding of the 11th European Conference on Object-Oriented Programming (ECOOP),* eds. M. Aksit and S. Matsuoka, 220–42. Berlin, Germany: Springer-Verlag.

[6] Arpaia, P., M. Buzio, L. Fiscarelli, and V. Inglese. November 2012. "A Software Framework for Developing Measurement Applications Under Variable Requirements." *AIP Review of Scientific Instruments* 83, no. 11, pp. 115103. doi: 10.1063/1.4764664

[7] Postolache, O., J.M.D. Pereira, M. Cretu, and P.S. Girao. 1998. "An Ann Fault Detection Procedure Applied in Virtual Measurement Systems Case." In *Proceeding of IEEE Instrumentation and Measurement Technology Conference, IMTC/98,* Vol. 1. St. Paul, MN: IEEE.

[8] Catelani, M., and S. Giraldi. March 1999. "A Measurement System for Fault Detection and Fault Isolation of Analog Circuits." *Measurement* 25, no. 2, pp. 115–22. doi: http://dx.doi.org/10.1016/s0263-2241(98)00072-4

[9] Arpaia, P., G. Lucariello, and A. Zanesco. October 2007. "Automatic Fault Isolation by Cultural Algorithms with Differential Influence." *IEEE Transactions on Instrumentation and Measurement* 56, no. 5, pp. 1573–82. doi: http://dx.doi.org/10.1109/tim.2007.903604

[10] Arpaia, P., M.L. Bernardi, G. Di Lucca, V. Inglese, and G. Spiezia. 2010. "An Aspect-Oriented Programming-Based Approach to Software Development for Fault Detection in Measurement Systems." *Computer Standards & Interfaces* 32, no. 3, pp. 141–52. doi: http://dx.doi.org/10.1016/j.csi.2009.11.009

[11] Arpaia, P., L. Fiscarelli, G. La Commara, and F. Romano. January 2011. "A Petri Net-Based Software Synchronizer for Automatic Measurement Systems." *Instrumentation and Measurement, IEEE Transactions* 60, no. 1, pp. 319–28. doi: http://dx.doi.org/10.1109/tim.2010.2046602

[12] Gupta, R.K., C.N. Coelho, and G. De Micheli. 1992. "Synthesis and Simulation of Digital Systems Containing Interacting Hardware and Software Components." *29th ACM/IEEE Design Automation Conference*, pp. 225–30. Anaheim, CA: IEEE.

[13] von Praun, C., H.W. Cain, J. Choi, and K.D. Ryu. 2006. "Conditional Memory Ordering." *Proceeding of the 33th International Symposium on Computer Architecture (ISCA)*, Vol. 34, pp. 41–52. Washington, DC: IEEE Computer Society.

[14] Arpaia, P., M.L. Bernardi, G. Di Lucca, V. Inglese, and G. Spiezia. May 2008. "Aspect Oriented-based Software Synchronization in Automatic Measurement Systems." *Instrumentation and Measurement Technology Conference Proceedings, 2008 (IMTC 2008)*, pp. 1718–21. Victoria, BC: IEEE.

[15] Birrell, A., J. Guttag, J. Horning, and R. Levin. November 8–11, 1987. "Synchronization Primitives for a Multiprocessor: A Formal Specification." In *Proceeding of the 11th ACM Symposium Operating System Principles*, 94–102. Austin, TX: ACM Press.

[16] Anderson, T.E., B.N. Bershad, E.D. Lazowska, and H.M. Levy. February 1992. "Scheduler Activations: Effective Kernel Support for the User-Level Management of Parallelism." *ACM Transactions On Computer Systems (TOCS)* 10, no. 1, pp. 53–79. doi: http://dx.doi.org/10.1145/121132.121151

[17] Murata, T. April 1989. "Petri Nets: Properties, Analysis and Applications." In *Proceedings of IEEE* 77, no. 4, pp. 541–80. doi: http://dx.doi.org/10.1109/5.24143

[18] Bosch, J. 1999. "Design of an Object-Oriented Framework for Measurement Systems." In *Domain-Specific Application Frameworks*, eds. M. Fayad, D. Schmidt, and R. Johnson, pp. 177–205. ISBN 0-471-33280-1. New York: John Wiley.

[19] Stoyanov, B., S. Stefanov, J. Beyazov, and V. Peichev. 2006. "Contemporary Methods and Devices for Automatic Measurement." *Problems of Engineering Cybernetics and Robotics* 57, pp. 79–86.

[20] Plone Fundation, 2014. Welcome to the TANGO Controls website. [Online]. http://www.tango-controls.org/

[21] Nogiec, J.M., J. Di Marco, S. Kotelnikov, K. Trombly-Freytag, D.Walbridge, and M. Tartaglia. June 2006. "Configurable Component-Based Software System for Magnetic Field Measurements." *IEEE Transactions on Applied Superconductivity* 16, no. 2, pp. 1382–5. doi: http://dx.doi.org/10.1109/tasc.2005.869672

[22] Ding, L., and Y. Shen. May 5–7, 2009. "Real Time Performance Analysis and Evaluation of CAN Bus with an Extended Petri Net Model." In *Proceeding of the IEEE I2MTC*, 1081–84. Singapore: IEEE.

[23] Frankowiak, M.R., R.I. Grosvenor, and P.W. Prickett. June 1 2005. "A Petri-Net Based Distributed Monitoring System Using Pic Microcontrollers." *Microprocessors and Microsystems* 29, no. 5, pp. 189–96. doi: http://dx.doi.org/10.1016/j.micpro.2004.08.003

[24] Calderaro, V., V. Galdi, A. Piccolo, and P. Siano. August 19–24, 2007. "DG and Protection Systems in distribution network: Failure Monitoring System Based on Petri Nets." In *Proceeding of the Bulk Power System Dynamics and Control–VII. Revitalizing Operational Reliability, 2007 iREP Symposium*, pp. 1–7. Charleston, SC: IEEE

[25] Hadjicostis, C.N., and G.C. Verghese. September 2000. "Power System Monitoring Using Petri Net Embeddings." In *Proceeding Institute of Electrical Engineering—Generation, Transmission and Distribution* 147, no. 5, pp. 299–303. doi: http://dx.doi.org/10.1049/ip-gtd:20000657

[26] Huiqin, Z., G. Jun, X. Youbao, and L. Wei. August 16–18, 2007. "Modeling and Analysis of a Testing System Using Hybrid Petri Net." In *Proceeding of the 8th ICEMI*, 1465–70. Xi'an, China: IEEE.

[27] Xiaoli, W., C. Guangju, X. Yue, and G. Zhaoxin. August 16–18, 2007. "Fault Detection and Diagnosis Based on Time Petri Net." In *Proceeding of the 8th ICEMI*, 3259–63. Xi'an, China: IEEE.

[28] Lindner, G., M. Heiner, and T. Kobienia. October 14–17, 1996. "Deadlock Detection in a Distributed Implementation of a Visualization System for Medical Measurement Signals." In *Proceeding of the IEEE International Conference on Systems, Man, and Cybernetics*, Vol. 3, pp. 2299–04. Beijing, China.

[29] Lukaszewski, R., and W. Winiecki. May 2008. "Petri Nets in Measuring Systems Design." *IEEE Transactions on Instrumentation and Measurement* 57, no. 5, 952–62. doi: http://dx.doi.org/10.1109/imtc.2006.328678

[30] Bilski, P., and R. Lukaszewski. September 6–7 2007. "Petri Nets Model of DAQ Block in the Measurement System." In *Proceedings at the 4th IEEE Workshop on Intelligent Data Acquisition and Advanced Computing Systems: Technology and Applications*, pp. 268–73. Dortmund, Germany.

[31] Papelis, Y.E., and T.L. Casavant. March 1992. "Specification and Analysis of Parallel/Distributed Software and Systems by Petri Nets with Transition Enabling Functions." *IEEE Transactions on Software Engineering* 18, no. 3, pp. 252–61. doi: http://dx.doi.org/10.1109/32.126774

[32] Woodside, M. 2000. "Software Performance Evaluation by Models." In *Performance Evaluation: Origins and Directions*, eds. G. Haring, C. Lindemann, and M. Reiser. Berlin, Germany: Springer-Verlag.

[33] Topic, G., D. Jevtic, and M. Kunstic. 2008. "Petri Net-Based Simulation and Analysis of the Software Development Process." In *Knowledge-Based Intelligent Information and Engineering Systems*, eds. I. Lovrek, R.J. Howlett, and L.C. Jain. Berlin, Germany: Springer-Verlag.

[34] Zhou, M., and K. Venkatesh. 1999. *Modeling, Simulation, and Control of Flexible Manufacturing Systems: A Petri Net Approach*. Singapore: World Scientific, ser. Series in Intelligent Control and Intelligent Automation.

[35] Kumanan, S., and K. Raja. 2008. "Modeling and Simulation of Projects with Petri Nets." *American Journal of Applied Sciences* 5, no. 12, pp. 1742–49. doi: http://dx.doi.org/10.3844/ajassp.2008.1742.1749

[36] Billington, J., M. Diaz, and G. Rozenberg. 1999. *Application of Petri Nets to Communication Networks*. New York: Springer-Verlag.

[37] Cortadella, J., M. Kishinevsky, A. Kondratyev, L. Lavagno, and A. Yakovlev. 1997. "Petrify: A Tool for Manipulating Concurrent Specifications and Synthesis of Asynchronous Controllers." *IEICE Transactions on Information and Systems*, Vol. E80-D, no. 3, pp. 315–25.

[38] Arpaia, P., L. Fiscarelli, G. La Commara, and C. Petrone. December 2011. "A Model-Driven Domain-Specific Scripting Language for Measurement-System Frameworks." *IEEE Transactions on Instrumentation and Measurement* 60, no. 12, pp. 3756–66. doi: http://dx.doi.org/10.1109/tim.2011.2149310

[39] Product Information: "What is NI LabVIEW?" January 2011. http://www.ni.com/labview/whatis/

[40] G Programming Reference Manual. January 2011. http://www.ni.com/pdf/manuals/321296b.pdf

[41] Scripting Languages and NI LabVIEW. January 2011. ftp://ftp.ni.com/pub/devzone/pdf/tut_7671.pdf

[42] Pakin, S. October 2007. "The Design and Implementation of a Domain-Specific Language for Network Performance Testing." *IEEE Transactions on Parallel and Distributed Systems* 18, no. 10, pp. 1436–49. doi: http://dx.doi.org/10.1109/tpds.2007.1065

[43] Bennett, M., R. Borgen, K. Havelund, M. Ingham, and D. Wagner, November 2010. "Prototyping a Domain-Specific Language for Monitor and Control Systems." *Journal of Aerospace Computing, Information, and Communication* 7, no. 11, pp. 338–64. doi: http://dx.doi.org/10.2514/1.40331

[44] Schmidt, D.C. February 2006. *Model-Driven Engineering*. IEEE Computer Society 39, no. 2.

[45] Arpaia, P., M. Buzio, L. Fiscarelli, V. Inglese, G. La Commara, and L. Walckiers. May 2009. "Measurement-Domain Specific Language for Magnetic Test Specifications at CERN." In *Proceedings of IEEE Conference on Instrumentation and Measurement Technology*, pp. 1716–20. Singapore: IMT.

[46] Spinellis, D. February 2001. "Notable Design Patterns for Domain-Specific Languages." *Journal of Systems and Software* 56, no. 1, pp. 91–9. doi: http://dx.doi.org/10.1016/s0164-1212(00)00089-3

[47] Bosch, J., and G. Hedin. April 1996. *Proceedings of Workshop Compiler Techniques for Application Domain Languages and Extensible Language Models*, pp. 96–173. Lund University, Lund, Sweden: Lund University, Department of Computer Science.

[48] Chapman, N.P. January 1998. *LR Parsing: Theory and Practice*. 1st ed. Cambridge, U.K.: Cambridge Univ. Press.

[49] OMG Mission Statement. January 2011. http://www.omg.org/gettingstarted/gettingstartedindex.htm

[50] Arpaia, P., L. Fiscarelli, and G. La Commara. 2010. "Advanced User Interface Generation in the Software Framework for Magnetic Measurements at CERN." *Metrology and Measurement Systems* 17, no. 1, pp. 27–38. doi: http://dx.doi.org/10.2478/v10178-010-0003-y

[51] Mayers, B., S.E. Hudson, and R. Pausch. 2000. "Past, Present and Future of User Interface Software Tools." *ACM Transactions on Computer-Human Interaction* 7, no. 1, pp. 3–28. doi: http://dx.doi.org/10.1145/344949.344959.

[52] Beaudouin-Lafon, M. 2005. "Interactions as First-class Objects." In *Proceedings of the ACM CHI 2005 Workshop on the Future of User Interface Design Tools*. ACM Press.

[53] Hayes, P.J., P. Szekely, and A. Richard. April 14–18, 1985. "Design Alternatives for User Interface Management Systems Based on Experience with COUSIN." *CHI '85 Proceedings of the SIGCHI Conference on Human Factors in Computing Systems*, pp. 169–75. San Francisco, CA: ACM.

[54] Schulert, A.J., G.T. Rogers, and J.A. Hamilton. April 1985. "ADM-A Dialogue Manager," *CHI '85 Proceedings of the SIGCHI Conference on Human Factors in Computing Systems*, pp. 177–83. San Francisco, CA: ACM.

[55] Olsen, D.R. 1986. "Mike: The Menu Interaction Kontrol Environment." *ACM Transactions on Graphics* 5, no. 4, pp. 318–44. doi: http://dx.doi.org/10.1145/27623.28868

[56] Vander Zanden, B., and B.A. Myers. April 1–5 1990. "Automatic, Look-and-Feel Independent Dialog Creation for Graphical User Interfaces." In *CHI '90 Proceedings of the Conference on Human Factors in Computing Systems*, pp. 27–34. Seattle, WA: ACM.

[57] Sukaviriya, P., J.D. Foley, and T. Griffith. 1993. "A Second Generation User Interface Design Environment: The Model and The Runtime Architecture." In *Proceedings of the INTERACT '93 and CHI '93 Conference on Human Factors in Computing Systems*, pp. 375–82. Amsterdam, the Netherlands.

[58] Wiecha, C., W. Bennett, S. Boies, J. Gould, and S. Greene. 1990. "ITS: A Tool for Rapidly Developing Interactive Applications." *ACM Transactions on Information Systems* 8, no. 3, pp. 204−36. doi: http://dx.doi.org/10.1145/98188.98194

[59] Szekely, P., P. Luo, and R. Neches. 1993. "Beyond Interface Builders: Model-Based Interface Tools." In *Proceedings on CHI '93 Conference on Human Factors in Computing Systems*, pp. 383−90. Amsterdam, The Netherlands.

[60] Palanque, P., and F. Paternò. 1997. "Using Declarative Descriptions to Model User Interfaces With MASTERMIND." *Formal Methods in Human Computer Interactions*. Berlin, Germany: Springer-Verlag.

[61] Stirewalt, K., and S. Rugaber, 1998. "Automating UI Generation by Model Composition." Submitted to *Automated Software Engineering*, ASE '98, 13th IEEE International Conference.

[62] Lloyd, J.W. 1994. "Practical advantages of declarative programming." *Joint Conference on Declarative Programming*, GULP-PRODE, Vol. 94.

[63] Arpaia, P., M. Buzio, L. Fiscarelli, V. Inglese, and G. La Commara. September 2009. "Automatically-Generated User Interfaces for Measurement Software Frameworks: A Case Study on Magnetic Permeability at CERN." *XIX IMEKO World Congress, Fundamental and Applied Metrology*. Lisbon, Portugal: Curran Associates.

[64] Krasner, G., and S. Pope. August 1988. "A Cookbook for Using the Model-View-Controller User Interface Paradigm in Smalltalk-80." *Journal of Object-Oriented Programming* 1, no. 3, pp. 26−49.

[65] Abrams, M., C. Phanouriou, A.L. Batongbacal, S. Williams, and J.E. Shuster. 1999. "UIML: An Appliance- Independent XML User Interface Language." In *Proceedings of the Eighth International WWW Conference*. Toronto, Canada: Elsevier North-Holland.

[66] Achten, P., M. van Eekelen, and R. Plasmeijer. 2004. "Compositional Model-Views with Generic Graphical User Interfaces." In *Practical Aspects of Declarative Programming*, PADL04, vol. 3057 of LNCS. Edinburgh, UK: Springer.

[67] Achten, P., M. van Eekelen, and R. Plasmeijer. 2003. "Generic Graphical User Interfaces." Selected Papers of the 15th Int. *Workshop on the Implementation of Functional Languages*, IFL03, vol. 145 of LNCS. Edinburgh, UK: Springer.

[68] Lutteroth, C., and G. Weber. 2008. "Modular Specification of GUI Layout Using Constraints." In *Proceedings at ASWEC 2008—19th Australian Conference on Software Engineering*, IEEE Press.

CHAPTER 5

QUALITY ASSESSMENT OF MEASUREMENT SOFTWARE

Men acquire a particular quality by constantly acting a particular way...
you become just by performing just actions, temperate by performing temperate actions,
brave by performing brave actions.

—Aristotle

5.1 OVERVIEW

In this chapter, the assessment of the software quality for measurement software frameworks is presented. First, main concepts of software quality are introduced very synthetically from a general perspective. Then, the approach proposed in the standard ISO 9126 is chosen as a reference model for the quality assessment of measurement and test software. In particular, a practical approach, the *quality pyramid*, to software quality is presented for (a) characterizing the design of an Object-Oriented project, (b) finding possible problems, and (c) defining the related corrective actions. Finally, a method based on specific metrics for assessing the degree of flexibility achieved by a software framework for measurement applications is presented.

5.2 SOFTWARE QUALITY

Quality is a key issue in professional software development. In general, quality can be defined in two, not-exclusive, ways, as the degree to which a system, component, or process meets: (a) specified technical working requirements and (b) customer or final user needs or expectations [1]. This definition offers the corresponding two most common interpretations of the word "quality" at (a) an engineering level, as conformance to requirements (*inner quality*) and (b) a marketing level, as measure of user satisfaction (*outer quality*).

Software quality can be approached from different perspectives [2–4], but a specific common definition is provided as "the capability of a software product to satisfy stated and implied needs when used under specified conditions" [5]. Pursuing software quality is always worthwhile, because the cost of achieving a high quality level is widely overtaken by the cost of nonquality (i.e., of having software unable of providing the required features when needed).

5.2.1 SOFTWARE QUALITY METRICS

Software quality cannot be assessed without a clear definition of a method for its measurement in an objective way. For this reason, as a first step, several metrics have been introduced for this aim. The term metric is defined as a "measure of the degree to which a process or product possesses a certain quality characteristic" [6]. In this chapter, only the metrics most suitable for assessing the quality of software frameworks for measurement applications are discussed. They are listed in Tables 5.1 and 5.2, by providing a brief description for each of them.

For an objective assessment, when using metrics the two major issues are (a) to define a suitable reference and (b) above all, for this reference, to determine proper levels (*quality thresholds*). The main objective is to not look for thresholds perfect from a theoretical viewpoint, but for values turning out to be useful in a practical perspective, in order to detect possible software artefacts. Two major sources of useful threshold levels can be identified [7]:

- *Probability information*, leading to thresholds based on statistical significance. One or more reference points are used to split numerical spaces into meaningful intervals (acceptance or rejection regions in statistical decision making). By applying simple statistical techniques to the data collected for each metric, the average (AVG) can be used to estimate the typical values, and the standard deviation (STD) to define higher/lower margins of the confidence interval as AVG ± STD.

Table 5.1. Metrics catalog [8]

Metric	Description
Cyclomatic complexity (CYCLO)	Logic complexity of a software module, as the number of linearly independent paths.
Essential complexity (ESS)	Cyclomatic complexity after replacing all well-structured control structures by a single statement.
Class depending child (CDC)	Class depending at least on one of its children.
Class depth (DEPTH)	Depth of a class within the inheritance hierarchy.
Multiple inheritance (FAN IN)	Number of immediate base classes.
Response for class (RFC)	Number of methods, including inherited ones.
Coupling between objects (CBO)	Number of other classes using a type, data, or member from that class (coupling). All the couplings to a given class counts as one toward the metric total.
Lack of cohesion of methods (LOCM/LCOM)	Cohesion between class data and methods.
Weighted methods for class (WMC)	Sum of cyclomatic complexity of all nested functions or methods.
Access to foreign data (ATFD)	Number of attributes from unrelated classes accessed directly or through accessory methods.
Changing classes (CC)	Number of classes in which the methods calling the measured method are defined.
Coupling intensity (CINT)	Number of distinct operations called by the measured operation.
Changing methods (CM)	Number of distinct methods that call the measured method.
Lines of code (LOC)	The number of lines that contains source code.
Tight class cohesion (TCC)	Relative number of method pairs of a class that access in common at least one attribute of the measured class.
Weight of a class (WOC)	Number of public methods divided by the total number of public members.

(Continued)

Table 5.1. (*Continued*)

Metric	Description
Number of accessory methods (NOAM)	Number of accessory (getter and setter) methods of a class.
Number of public attributes (NOPA)	Number of public attributes of a class.
Number of accessed variables (NOAV)	Number of variable accessed directly by the measured operation.
Locality of attribute accesses (LAA)	Number of attribute from the method definition class, divided by the total number of variable accessed.
Foreign data providers (FDP)	Number of classes in which the attributes accessed are defined.
Maximum nesting level (MAXNESTING)	Maximum nesting level of control structures within an operation.

- *Generally accepted semantics*, leading to thresholds based on information considered as common, widely accepted knowledge. This knowledge could be in turn based on former statistical observations, but in general their values have become, to some extent, part of our culture and can be inferred without a statistical assessment.

5.2.2 SOFTWARE QUALITY MODELS

The correct use of metrics is tuned, and correspondingly their misuse is avoided, by assessing them within the frame of a quality model. A model is an abstraction of reality, allowing useless details to be discarded and entities or concepts to be viewed from a particular perspective [9], by understanding simultaneously the interactions among the parts forming the whole system of interest. The quality of a system is the result of the quality of its elements and their interactions. A model can be used to predict or assess the quality, the latter being the aim of this chapter.

A considerable amount of work has been devoted to the formulation of the so-called quality models. One of the first was proposed by Gilb [10], according to whom any quality characteristic can be measured directly. The quality concept is broken into component parts until each can be stated in terms of directly measurable attributes. Other models were proposed by Bohm [6] and McCall [11]. These are hierarchical models, based on the

Table 5.2. Metrics catalogue (NDD, AHH, DOF, and DOS are proportions)

Metric	Description
Coupling dispersion (CDISP)	Number of classes in which the operation called from the measured operation are defined, divided by CINT.
Number of packages (NOP)	Number of high level packages (e.g., packages in Java or namespaces in C++).
Number of classes (NOC)	Number of classes defined in the software system, not including library classes.
Number of operations (NOM)	Number of user-defined operations (methods and global functions).
Number of operation calls (CALL)	Number of distinct operation calls (invocations) made by all the user-defined operations.
Number of called classes (FOUT)	Sum of the classes whose operations call methods, for all the operations defined by the user.
Number of direct descendants (NDD)	Average of subclasses for each class tells to what extent abstractions are refined by means of inheritance. Interfaces are not counted.
Height of inheritance tree (HIT)	Average number of inheritance levels in a class hierarchy.
Average hierarchy height (AHH)	Average length of the paths from a root class to its farthest subclasses, measures the deepness of the class hierarchy. Interfaces are not counted.
Degree of focus (DOF)	Measure of the level of dedication of a component to every concern in the system.
Degree of scattering (DOS)	Measure of the level of scattering of a concern within all the modules in the systems.

assumption of the existence of several important high-level quality factors determined by lower level criteria, supposed much easier to measure than the corresponding factors. Actual measures, the metrics, are proposed for the criteria. The model describes all the relationships between factors and criteria, so that the former can be quantified in terms of measures of their dependent criteria. This conception of modeling quality was more recently the basis of international efforts that led to the development of a standard for software quality measurement, defining for the software (a) a quality model (ISO 9126 [12–15]), (b) a measurement process (ISO 15939 [16]), and (c) an evaluation process (ISO 14598 [17]). The standard ISO 9126-1 [12] recommends six quality characteristics, further refined in subcharacteristics, as the basic set for quality evaluation. Furthermore, the standards ISO 9126-2, -3, and -4 [13–15] define metrics for measuring characteristics and subcharacteristics. Anyway, the metric list is not finalized and no clear indications are provided about its mapping to the quality characteristics. Given a particular problem, techniques like the Goal-Question-Metric (GQM) [18] can help in identifying which measures are to be taken into account to monitor and improve quality in the specific case.

5.3   THE STANDARD ISO 9126

A definition of software quality, along with guidance for its assessment, is provided by international standards [12–17]. This section aims at assessing the level of quality achievable by a software framework for measurement applications, according to the guidelines set by these standards.

In particular, the software quality model provided by the standard ISO 9126 defines six quality characteristics (Figure 5.1) [12]:

- Functionality: "The capability of the software product to provide functions meeting stated and implicit needs, when the software is used under specified conditions."
- Reliability: "The capability of the software product to maintain a specified level of performance when used under specified conditions."
- Efficiency: "The capability of the software product to provide appropriate performance, relative to the amount of resources used, under stated conditions."
- Usability: "The capability of the software product to be understood, learned, used, and attractive to the user, when used under specified conditions."
- Portability: "The capability of the software product to be transferred from one environment to another."
- Maintainability: "The capability of the software product to be modified. Modifications may include corrections, improvements, or adaptation of the software to changes in environment, and in requirements and functional specifications."

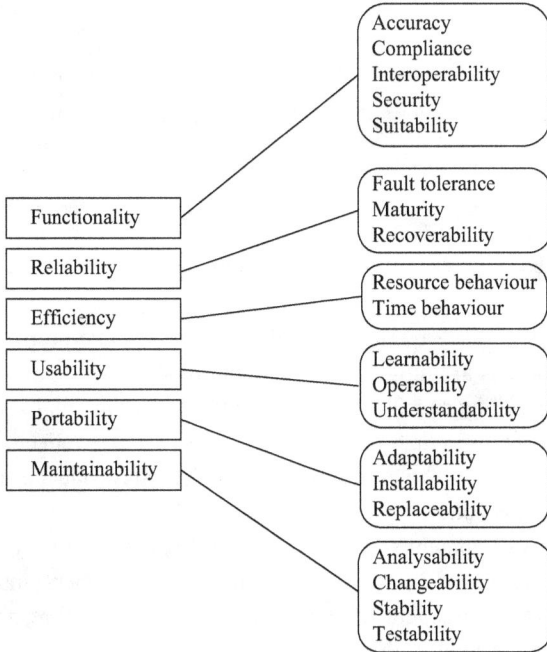

Figure 5.1. The ISO 9126 quality model [12].

The standard defines also an additional quality characteristic:

- Quality in use: The capability of the software product to enable specified users to "achieve specified goals with effectiveness, productivity, safety, and satisfaction in specified contexts of use."

The quality characteristics have well-defined subcharacteristics, and the standard allows the user to define his or her own sub-subcharacteristics according to a hierarchical structure. The ISO framework is completely hierarchical; each subcharacteristic is related to only one characteristic.

The corresponding quality model defines three categories of quality:

i. Software quality in use, related to the software application inside its operational environment, for carrying out certain tasks by specific users.

ii. External software quality, providing a black-box view of the software and addressing properties related to the execution of the software on a computer hardware and applying an operating system.

iii. Internal software quality, providing a white-box view and addressing properties of the software product typically available during the development.

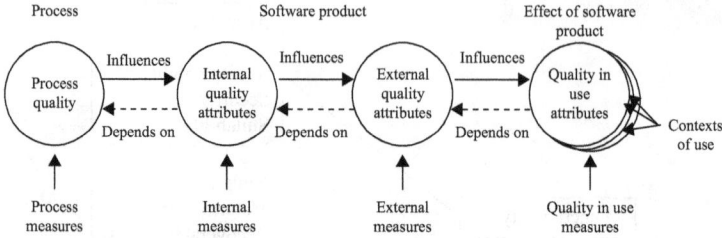

Figure 5.2. Approaches to software quality according to ISO 9126 [12].

Internal software quality is mainly related to static properties of the software and has an impact on its external quality, which in turn affects quality in use (Figure 5.2). As shown in the figure, internal and external attributes refer to product quality, while quality in use reflects the user's view of the product. The process through which this product is obtained is not taken into account.

At the development stage of a software framework, the system is not yet widely employed by users others than the developers, thus making premature the assessment of the quality in use. Therefore, the quality assessment presented in the following sections is dedicated to a finished product. More technically, the properties of the software's design and code (internal quality) are analyzed at a static level by means of purely internal metrics [14].

5.4 QUALITY PYRAMID

A more practical approach to software quality is presented in [7], for (a) characterizing the design of the Object-Oriented part of a software framework, (b) finding the possible problems, and (c) proposing the corresponding corrective actions. According to this approach, some design metrics [19], namely, capable of capturing the quality of the design at a certain point in the software development cycle, are used in the frame of the Goal-Question-Metric (GQM)[1] technique [18] to effectively characterize and evaluate the design of an Object-Oriented system.

As a matter of fact, the quality characterization of an Object-Oriented system is a complex task that cannot rely on the evaluation of single

---

[1]Usually the Goal-Question-Metric (GQM) model is described as a six-step process, where the first three objectives aim at fastening the identification of the right metrics, while the other three phases belong to the data obtained and specifically to how to process the actual measurement results for making effective decisions.

metrics in a nonorganized way. It requires choosing suitable metrics, computing their values, and above all correlating them in a proper manner in order to draw significant conclusions. Moreover, effective metrics have to reflect three main aspects (Figure 5.3) [7]: (a) size and complexity, to take into account how big and complex a system is; (b) coupling, to know to which extent classes are mutually coupled; and (c) inheritance, to assess how much and how well the concept of inheritance is used.

To better clarify and understand these three aspects, the Overview Pyramid was introduced in [7].

The pyramid is a metric-based means to both describe and characterize the overall structure of an Object-Oriented system by quantifying its complexity, coupling, and inheritance. It can be seen as "a graphical template for presenting and interpreting system-level measurements of software quality in a unitary manner" [7]. An overview pyramid is composed of three parts, corresponding to the aforementioned characteristics (Figure 5.4): (a) *system size and complexity* (filled in dark grey in Figure 5.4), (b) *system coupling* (in light grey in Figure 5.4), and (c) *system inheritance* (in white).

5.4.1 SYSTEM SIZE AND COMPLEXITY

The left bottom side of the pyramid (filled in dark grey in Figure 5.4) gathers information characterizing size and complexity of the software

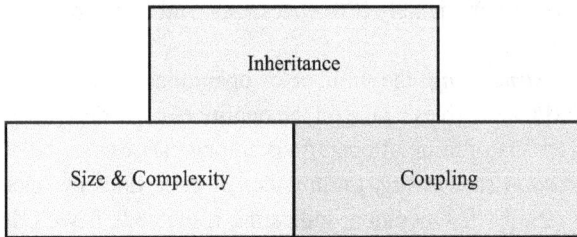

Figure 5.3. The three major aspects quantified by the overview pyramid [7].

					NDD	0.31			
					HIT	0.12			
			20.21	NOP	19			
		9.42	NOC		384			
	9.72	NOM			3618	NOM	4.18	
0.15	LOC				35175	15128	CALL	0.56
CYCLO					5579	8590		FOUT

Figure 5.4. Example of a completed overview pyramid (the metrics values refer to an application example).

system, provided by metrics directly computed on the source code. These simple and widely used metrics refer to the most significant modularity unit of an Object-Oriented system, in sequence from the highest (packages or namespaces) to the lowest level (lines of code, LOC). The pyramid provides each unit by a specific metric for measuring its quality. The considered metrics, placed one per line according to a top-down scheme,[2] are: the number of packages (NOP), the number of classes (NOC), the number of operations (NOM), the LOC, and the cyclomatic complexity (CYCLO) (Tables 5.1 and 5.2). From these basic absolute metrics, some proportions among the direct metrics (the values on the left in Figure 5.4) are computed by dividing the value of a metric by the next upper one. For example, the value of the proportion in the bottom left corner is obtained as ratio of the CYCLO and the LOC metrics and is equal to 0.15.

These proportions, unlike the direct metrics, have two important characteristics: (a) they are mutually independent, thus each value provides a distinct measure of a specific characteristic of the code organization, and (b) they simplify the comparison with other projects, because they are defined as ratios of absolute values and are therefore independent of the project size.

From this part of the overview pyramid, the further following proportions can be assessed about:

i. *high-level structuring*, the number of classes by package, computed as the ratio NOC/package, provides a clue on the packaging level of the system, namely if the packages tend to be coarse-grained or fine-grain;

ii. *class structuring*, the number of operations by class, as the ratio NOM/Class, gives a clue on the quality of class design, by measuring the distribution of operations among classes;

iii. *operation structuring*, the number of code lines by operation, as the ratio LOC/Operation, indicating if the code is well distributed among operations, that is, if the procedural programming is well structured;

iv. *intrinsic operation complexity*, the number of linearly-independent paths (cyclomatic complexity) by code line, assessed as ratio CYCLO/Code line, characterizing the conditional complexity found in the operations (density of branches with respect to the lines).

---

[2]In the pyramid, only the lines of code belonging to methods are counted.

5.4.2 SYSTEM COUPLING

The right part of the overview pyramid (filled in light grey in Figure 5.4) concerns the level of coupling in the software system, presented in terms of operation invocations. Two direct metrics are used to establish how the coupling is intensive and dispersed in the system (Tables 5.1 and 5.2): (a) CALL, that is, the number of distinct operation calls (invocations) made by all the user-defined operations, and (b) FOUT, that is, the sum of the classes whose operations call methods, for all the operations defined by the user. Also in this case, from these basic absolute metrics, proportions among the direct metrics (again reported in Figure 5.4) are computed by dividing the value of a metric by the next upper one.

Although these metrics describe the total amount of coupling of a software system, they are difficult to use for characterizing the system. Alternatively, two proportions can be calculated using the number of operations (NOM):

i. *Coupling intensity* (CALL/Operation), denoting the extent of coupling (collaboration) between the operations. This proportion tells, on average, the number of distinct operations called from each operation. High values indicate excessive coupling among operations in the code, that is, that the collaboration among operations is not well designed.

ii. *Coupling dispersion* (FOUT/Operation call), indicating how many classes are involved in the coupling, namely the sum of the classes whose operations call methods, for all the user-defined operations, by operation call.

5.4.3 SYSTEM INHERITANCE

The upper part of the overview pyramid (filled in white in Figure 5.4) is composed of two proportion metrics that globally characterize the use of inheritance. They give a clue of the extent to which some typical Object-Oriented features (generalization and polymorphism) are used. The two proportions are NDD, that is, the average of subclasses for each class, and HIT, that is, the average number of inheritance levels in a class hierarchy (Table 5.2). These two metrics apply to class hierarchies and characterize their shape by capturing two important complementary aspects: their width and height.

5.4.4 INTERPRETING THE PYRAMID

According to this approach, for characterizing an Object-Oriented system, the aforementioned eight proportions have to be computed. The proportions are preferred to the absolute metrics because they are independent of the project size, thus making their assessment easier. Moreover, they allow the use of thresholds based on statistical measurements, as explained in Section 5.2.1. In particular, metrics collected from a statistical base of Java and C++ projects are highlighted in [7].

5.5 MEASURING FLEXIBILITY

Classic and contemporary literatures in software design recognize the central role of flexibility. Moreover, one of the most significant attributes of software quality is the flexibility. Flexibility is maximized by means of structured design, modular design, Object-Oriented design, software architecture, design patterns, and component-based software engineering, among others.

During its life cycle, a flexible software system is forced to face variable requirements. As a consequence, the implementation has to be adapted to provide a solution to problems in new application domains. An evolution step is defined as the unit of evolution referred to a particular change in the implementation.

It has been observed that predicting the class of changes is the key to understanding software flexibility. During the phases of design and development of the software, initially the changes that are likely to occur over the lifetime of the product are characterized. Since it is impossible to predict the actual changes, the predictions will refer to classes of possible changes [20].

The notion of evolution step can be used for estimating software flexibility [21]: Software is more flexible than a given value b toward a particular evolution step, if the number of changes required for the software upgrade is smaller than the number of changes required for b. Thus, the complexity of an evolution step measures how inflexible the implementation is toward a particular class of changes: Fewer changes are needed, the more flexible it is.

Therefore, it is useful to organize the software so that the items that are most likely to change are confined to a small amount of code, thus only this limited portion would be affected [20]. In other words, flexibility (measured in terms of the cost of the evolution process) is directly linked to the amount of code affected by the changes required in a particular evolution phase. Thus, a first approximation to measuring the cost of an evolution step ε is given by the evolution cost metric counting the number of modules affected by ε. Under the assumption that the costs of adding, removing, or changing each modular unit commensurate, the evolution cost metric is computed as the number of modules added, removed, or adjusted as a result of the evolution. This number is obtained by calculating the symmetric set difference between the sets of classes in the old (i_{old}) versus the adjusted $(i_{adjusted})$ implementations.

Formally [21]:

$$C_{classes}(\varepsilon) = \left| \left(Classes(i_{old}) - Classes(i_{adjusted}) \right) \cup \left(Classes(i_{adjusted}) - Classes(i_{old}) \right) \right| \quad (5.1)$$

This evolution cost metric is inadequate in some situations: When the evolution of different modules do not commensurate, the modules are not implemented yet, and the programming language does not support classes at all or adds other programming units (such as in the case of the Aspect-Oriented programming). Therefore, the metric is to be accommodated for varying degrees of modular granularity, as well as information on each module. This leads to the definition of the generalized cost metric [21, 22]:

$$C_{Modules}^{\mu}(\varepsilon) = \sum_{m \in \Delta Modules} (i_{old}, i_{adjusted}) \mu(m) \quad (5.2)$$

where $\Delta Modules$ $(i_{old}, i_{adjusted})$ is the symmetric set difference between the set of modules in i_{old} and the set of modules in $i_{adjusted}$.

The generalized metric is parameterized by the variables *Modules* and μ:

- The variable *Modules* represents any notion of module appropriate for the circumstances, such as class, procedure, method, aspect, and package [22];
- μ represents any software complexity metric meaningful in relation to a particular module m [22].

Finally, evolution complexity is a measure of growth, not an absolute value, and therefore it does not measure the actual cost of the evolution process, but rather how it grows.

REFERENCES

[1] IEEE Std 610-1990. 1991. IEEE Standard Computer Dictionary: A Compilation of IEEE Standard Computer Glossaries. (ANSI).

[2] Garvin, D. Fall 1984. "What Does "Product Quality" Really Mean?" In *Sloan Management Review*, pp. 25–45.

[3] Kitchenham, B., and S.L. Pfleeger. 1996. "Software Quality: The Elusive Target." *IEEE Software* 13, no. 1, pp. 12–21. doi: http://dx.doi.org/10.1109/52.476281

[4] Tervonen, I., and P. Kerola. 1998. "Towards deeper Co-Understanding of Software Quality." *Information and Software technology* 39, no. 14–15, 995–1003. doi: http://dx.doi.org/10.1016/s0950-5849(97)00060-8

[5] ISO. 1994. "ISO 8402: Quality Management and Quality Assurance—Vocabulary." *International Organization for Standardization*, 2nd ed. Geneva, Switzerland.

[6] Boehm, B.W., J.R. Brown, and M. Lipow. 1976 "Quantitative Evaluation of Software Quality," *Proceedings of the 2nd International Conference on Software engineering (ICSE '76)* pp. 592–605. San Francisco, CA, USA: IEEE Computer Society Press.

[7] Lanza, M., and R. Marinescu, 2006. *Object-Oriented Metrics in Practice*. Secaucus, NJ: Springer-Verlag New York, Inc.

[8] Scitools Understand. 2014. *Metrics*. [Online]: https://scitools.com/sup/metrics-2/.

[9] Fenton, N.E. 1991. *Software Metrics: A Rigorous Approach*. New York, NY: Chapman & Hall.

[10] Gilb, T. 1987. *Principals of Software Engineering Management*. Reading, MA: Addison-Wesley.

[11] McCall, J.A., P.K. Richards, and G.F. Walters. 1977. "Factors in Software Quality," volume 1, 2, and 3, US. Rome Air Development Center Reports NTIS AD/A-049 014, NTIS AD/A-049 015 and NTIS AD/A-049 016. Springfield, VA: U.S. Department of Commerce.

[12] International Standard ISO/IEC 9126-1. 2001. "Software Engineering—Product Quality—Part 1: Quality Model," International Organization for Standardization, International Electro technical Commission.

[13] International Standard ISO/IEC 9126-2. 2003. "Software Engineering—Product Quality—Part 2: External Metrics," International Organization for Standardization, International Electrotechnical Commission.

[14] International Standard ISO/IEC 9126-3. 2003. "Software Engineering—Product Quality—Part 3: Internal Metrics," International Organization for Standardization, International Electrotechnical Commission.

[15] International Standard ISO/IEC 9126-4. 2004. "Software Engineering—Product Quality—Part 4: Quality in Use Metrics," International Organization for Standardization, International Electrotechnical Commission.

[16] International Standard ISO/IEC 15939. 2002. "Software Engineering—Software Measurement Process," International Organization for Standardization, International Electrotechnical Commission.

[17] International Standard ISO/IEC 14598. 2000. "Software Engineering—Product Evaluation," International Organization for Standardization, International Electrotechnical Commission.

[18] van Solingen, R., and E. Berghout. 1999. *The Goal/Question/Metric Method: A Practical Guide for Quality Improvement of Software Development.* McGraw-Hill Inc., USA.

[19] Lorenz, M., and J. Kidd. 1994. *Object-Oriented Software Metrics: A Practical guide.* Upper Saddle River, NJ: Prentice-Hall.

[20] Parnas, D.L. May 1994. "Software Aging." In *Proceeding International Conference Software Engineering-ICSE.* Los Alamitos, CA: IEEE Computer Society Press, pp. 279–87.

[21] Eden, A.H., and T. Mens. June 2006. "Measuring Software Flexibility." *IEEE Proceedings Software* 153, no. 3, 113–25. doi: http://dx.doi.org/10.1049/ip-sen:20050045

[22] Arpaia, P. 2012. "A Software Framework for Developing Measurement Applications Under Variable Requirements." *Review of Scientific Instruments* 83, no. 11, 115103. doi: http://dx.doi.org/10.1063/1.4764664

PART III

CASE STUDY

THE FLEXIBLE FRAMEWORK FOR MAGNETIC MEASUREMENTS AT CERN

Some of the greatest discoveries...consist mainly in the clearing away of psychological roadblocks which obstruct the approach to reality; which is why, post factum they appear so obvious.
—Arthur Koestler, The Sleepwalkers: A History of Man's Changing Vision of the Universe

6.1 OVERVIEW

In this chapter, an overview of a case study of the software framework for magnetic measurements realized at the European Organization for Nuclear Research (CERN) in cooperation with the University of Sannio, the Flexible Framework for Magnetic Measurements (FFMM) [1], is provided.

In the first part of the chapter, a background on the application context of testing magnets for particle accelerators is given. In particular, initially, the main magnetic measurement methods for magnets testing are examined. Subsequently, the automatic systems for magnetic measurements in use at the main research centers for high-energy physics are reviewed. Special attention is paid to the software package for measurement automation for magnets harmonic analysis, that is, the Magnetic Measurement Program (MMP), which was in use at CERN before the FFMM.

In the second part of the chapter, an outline of the FFMM project is given. In particular, primarily, the need for flexibility, arising from the experience of other measurement systems previously employed at CERN, is analyzed. Although the main topic of this book is the software for measurement control and data acquisition, FFMM is the main part of a wider project aimed at developing an entire platform for automating magnetic measurements, including new high-performance hardware. Therefore, a brief overview of main hardware components is provided, for the sake of completeness of the measurement automation approach. Then, the FFMM is presented by introducing its main design concepts and architecture. Subsequently, its main components, the *Fault Detector* and the *Synchronizer*, are described, by highlighting their architectures and the implementation of their classes. Finally, the application of the Measurement Domain Specific Language (MDSL) and the Advanced User Interface Generator to the magnetic measurements for magnet testing is reported.

6.2 METHODS FOR MAGNETIC FIELD MEASUREMENTS

The main objectives of the High-Energy Particle (HEP) accelerators are (a) to explore matter at a very small scale (down to 10^{-18} m), by means of radiations of wavelength smaller than the dimension to be resolved; (b) to produce new, massive particles in high-energy collisions (like the Higgs particle, discovered in July 2012); (c) to reproduce locally the very-high temperatures occurring in stars or in the early universe, and investigate nuclear matter in these extreme conditions; and (d) to exploit the electromagnetic radiation the hadrons emit when accelerated, particularly when the beam trajectory is curved by a magnetic field (centripetal acceleration), for medical application, as an example. CERN, one of the most important HEP laboratories, located at Geneva in Switzerland, was founded in 1953 to provide a deeper understanding of the matter and its contents. The last CERN achievement is the Large Hadron Collider (LHC), the biggest machine ever built by man: A circular accelerator that collides proton beams or heavier ions into lead and is housed in a 27-km long underground tunnel.

In the development of the LHC, superconductivity played a crucial role. It is a special phenomenon of zero electrical resistance and the expulsion of magnetic fields occurring in certain materials when cooled below a characteristic "critical" temperature, usually in the order of a few K, or tens of K (in this case, they are referred to as high-temperature superconductors) [2]. In the LHC design, superconductivity was exploited both to achieve high electrical current levels by reducing ohmic losses for magnetic conditioning of HEP beams and to keep compact the size of the machine in order to save capital cost and simultaneously reduce electrical power consumption. High-energy, high-intensity machines produce beams with *MJ* energy, so that conversion efficiency from the power grid to the beam must be maximized, by reducing ohmic losses. The beams circulating into the LHC are bent, focused, and corrected by electromagnets, many of them needing nominal currents of about 12 kA. In electromagnets, superconductivity suppresses ohmic losses, thus the only power consumption is related to the associated cryogenic refrigeration to achieve a critical temperature of a few K [3].

The coils of the LHC superconducting magnets are wound with cables of niobium–titanium (NbTi) (7,000 *km* in total),[1] working in superfluid helium either at 1.9 *K* or at 4.5 *K*. A vertical dipolar field B of 8.33 *T* is required to bend the proton beams, whereas the quadrupole magnets are designed for a gradient of 223 *Tm*$^{-1}$ and a peak field of about 7 *T*.

In storage rings like the LHC, stable beams have to run as long as possible on the circular orbit in order to increase the number of collisions between the counter-rotating beams. This imposes strong constraints on the tolerable field perturbations along the trajectory. Deviations from the dipole and quadrupole fields, even if short in both space and time, can induce instabilities reducing the life time of the beam. Higher-order multipoles correctors are required to compensate the unavoidable imperfections of dipole and quadrupole magnets.

The production of magnets, with high field quality, has been invariably assisted by various measurements, based on different methods depending on the goal and accuracy of the desired analysis.

The quantities of relevance for the magnetic field produced by accelerator magnets are the strength and direction of the produced field, the errors with respect to the ideal field profile (the so called *multipoles*), and the location of the magnetic center in case of gradient fields. For all the LHC magnets, the previous quantities are required as an integral or average over the magnet's length, because the global behavior of the magnet is desired for the beam conditioning.

In the following, an overview of the main methods for magnetic measurements is provided. Owing to their relevance in magnet testing for particle accelerators, in the following chapters of this book, more details are provided about the rotating coil method and the related harmonic analysis.

---

[1]In a LHC cable, each of the 6,000 to 9,000 superconducting filaments of NbTi is about 0.007 mm thick, that is, about 10 times thinner than a human hair. If all the filaments are added together, they would stretch to the Sun and back 6 times with enough left over for about 150 trips to the Moon [4].

6.2.1 ROTATING COILS

The rotating coil method [5, 6] is widely used for magnets with cylindrical bore, owing to its capability of measuring all the properties of the magnetic field (field strength, nonideality, angle, direction, and so on) integrated over the coil length with high precision. An induction coil is placed on a circular support and is rotated in the field (Figure 6.1). The coil's angular position is measured by an angular encoder, rigidly connected to the rotating support. The coil rotating in the field cuts the flux lines, and a voltage V is induced at the terminals. The voltage is integrated between two predefined angles φ by obtaining the flux change as a function of angular position.

6.2.1.1 Magnetic Field Harmonic Analysis

The LHC dipoles are 15-meters long with a beam aperture of 50 *mm* in diameter, giving indicating the possibility of considering the coils as infinitely long and in evaluating the magnetic field on the transversal plane *x-y* of the orthogonal radial section by neglecting the longitudinal component z. This 2-dimensional approximation is very convenient in describing the magnetic field B in terms of a complex variable z. In the central part of the dipole taking into account the properties of the analytical functions, it can be postulated that the generated magnetic field B can be expanded in the complex plane in a power series [7]:

$$B(z) = B_1 \sum_{n=1}^{\infty} \frac{C_n R_{ref}^{n-1}}{B_1} \left(\frac{z}{R_{ref}} \right)^{n-1} = B_1 \sum_{n=1}^{\infty} c_n \left(\frac{z}{R_{ref}} \right)^{n-1} \cdot 10^{-4} \qquad (6.1)$$

where C_n is in *units* of $T \cdot m^{l-n}$ while $c_n = C_n \frac{R_{ref}^{n-1}}{B_1}$ are the multipoles normalized to the main dipole field and referred to a reference radius R_{ref} (17 mm for LHC main dipoles test). In this way, all the series

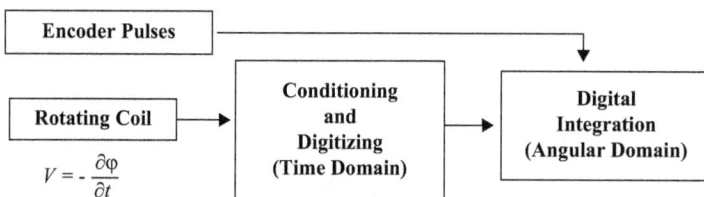

Figure 6.1. Rotating coils measurement principle.

coefficients c_n do not have dimension and are expressed in so-called *units* of the main field at the reference radius. They are then multiplied by a scaling factor (104 for LHC main dipoles test), namely, the order of the ratio between the main field and the field errors. In the complex plane, the coefficient C_n can be decomposed in its normal and skewed term, that is, real and imaginary part, respectively [8]:

$$C_n = B_n + iA_n \tag{6.2}$$

By using the aforementioned decomposition, and by applying the scaling factor to the normal and skew field components deduced from Equation 6.2, the field components in units of the main field B_1 can be expressed:

$$
\begin{cases}
a_n = A_n \dfrac{R_{ref}^{n-1}}{B_1} \cdot 10^4 \\[3mm]
b_n = B_n \dfrac{R_{ref}^{n-1}}{B_1} \cdot 10^4
\end{cases}
\tag{6.3}
$$

The existence of non-null b_n or a_n, or both, coefficients reflects the fact that the magnetic field generated by the superconducting coil, for example, in a dipole is not ideally dipolar, but conversely it is affected by higher-order components (multipoles, e.g., quadrupole, sextupole, and so on). The multipole components are generated by the difference between the ideal and the actual current distribution in the electromagnetic coil. All undesired multipole components other than the main field are referred to as *field errors*. They can be associated with the geometry approximation of the superconducting coils, but also they can have origins from the different elements and materials used in the magnet as a whole.

The rotating coil method eliminates the time dependence [9], that is, the influence of variations in the rotation speed. As a matter of fact, the field is integrated between two successive measurement angular points and the resulting value depends only on the initial and final points, and not on the time needed to achieve them, by relaxing requirements for uniform rotation greatly.

Differential measurements are beneficial to increase the resolution of high-order multipoles, several orders of magnitude smaller than the main field. This is realized by using a set of compensation coils mounted on the rotation support [10]. The signal from the compensation coils is used to suppress analogically the strong contribution from the main field. The compensated signal is analyzed in Fourier series together with the absolute signal of the outermost rotating coil in order to obtain the main field, as well as the high-order multipoles [8]. The overall uncertainty of the integral field strength and of the harmonics depends on the rotating shaft type. The new system developed at CERN can reach a bandwidth of harmonic measurement up to 100 *Hz* with a resolution of ±10 *ppm*.

6.2.2   STRETCHED WIRE

The stretched-wire technique is based on the induction method [11, 12]. A thin wire, with a diameter of 0.1 *mm*, is stretched in the magnet's bore between two precision stages. A motion results in a voltage at the two ends of the wire, whose integral is the magnetic flux through the area scanned by the motion. The method, a robust null technique with very-high resolution, provides a measurement of the integral field, of the field direction, and of the magnetic axis.

Metrological performance depends on the accuracy of the precision stages driving the wire motion (±1 μm), on the effectiveness of the sag correction, and on the alignment errors during installation. The overall uncertainty on the integrated strength and on the angle measurement was estimated as ±5 *units* and ±0.3 *mrad*, respectively [11, 12].

6.2.3   MAGNETIC RESONANCE TECHNIQUE

The nuclear magnetic resonance technique is considered as the primary static standard for calibration. It is frequently used, not only for calibration purposes, but also for high-accuracy field mapping. Based on an easy and accurate frequency measurement, it is independent of temperature variations. Commercially available instruments measure fields in the range from 0.011 *T* up to 13 *T*, with an accuracy better than ±10 *ppm*.

In practice, a sample of water is placed inside an excitation coil, powered from a radiofrequency oscillator. The precession frequency of the nuclei in the sample is measured either as nuclear induction (coupling into a detecting coil) or as resonance absorption [13]. The measured frequency is directly proportional to the strength of the magnetic field with coefficients of 42.57640 *MHz/T* for protons and 6.53569 *MHz/T* for deuterons.

Main advantages of the method are its very-high accuracy, its linearity, and the static operation of the system. The main disadvantage is the need for a rather homogeneous field in order to obtain a sufficiently coherent signal.

6.2.4 HALL PROBES

Hall probes exploit the Hall effect to measure magnetic fields [14]. When a current is flowing in a solid penetrated by a magnetic field, this field generates a voltage perpendicular to the current and the field itself. This voltage is large enough to be practical only for semiconductors [15]. The main uncertainty factor is due to the temperature coefficient of the Hall voltage.

The Hall probes permit the analysis of inhomogeneous fields because they measure the field locally. Conversely, the integral measurement over the entire magnet's length is more challenging, because the Hall sensors are quite small requiring either long and complex probes or many measurements steps.

6.3 AUTOMATIC SYSTEMS FOR MAGNETIC MEASUREMENTS

The need for automatic measurement systems applied to magnetic measurements buds from their complexity, in particular from the variety of tests employing various hardware setups and procedures. Usually, in accelerator systems, magnetic measurements require many measurement setups with different organization and technology. A commercial system to satisfy the need for automatic systems applied to magnetic measurements is not available on the market. For this reason, in the last few years, the main research centers have developed automatic systems integrating hardware (devices, instrumentation) with software (data acquisition, analysis, user interfacing).

At CERN for the LHC [16], the paradigm of the Front-End Software Architecture (FESA), provides a suitable PC front-end for interfacing the control instrumentation. The FESA infrastructure includes the following components: (a) an Object-Oriented real-time framework, to implement common functions with reusable components; (b) a graphical tool, namely a XML editor, in order to develop design, deployment, and instantiation schemas; (c) code generation; and (d) test environment. However, a strong collaboration and involvement at the lowest level of FESA is still required in order to adapt the architecture to the control requirements.

At the Fermi National Accelerator Laboratory, a new software system to test accelerator magnets was developed to be extensible to all the magnetic measurements, as well as to handle various types of hardware technologies and analysis algorithms [17]. The software is a configurable component-based framework, allowing for easy reconfiguration and runtime modification. Data acquisition, user interface, and analysis can be configured to create different measurement systems, tailored to specific requirements. Each test can be controlled and configured by a dedicated script.

Other subnuclear research centers (Alba, Soleil, Elettra, and ESRF) collaborated to develop a suitable software framework for testing accelerator magnets [18]. This Consortium proposes TANGO, an Object-Oriented system, to handle measurement applications. Tango is a distributed control system, where all the objects are representations of devices. The devices can be distributed or remotely interconnected. An archiving service stores data coming from the Tango control system into different types of databases.

6.4 SOFTWARE FOR MAGNETIC MEASUREMENTS AT CERN

Many magnetic measurement systems are currently used at CERN (rotating coils, stretched wire, and so on), and different software packages are employed for control, data acquisition, and analysis. These systems were developed incrementally during the years, without focusing specifically on their software quality, namely flexibility and reusability.

An example of such a software packages is the MMP [19], used in the past for the series tests of the LHC superconducting magnets. For its importance in the test activities carried out at CERN, mainly based on the rotating coil technique [20], and for its characteristic representative of the previous generation of control and acquisition systems, more details about MMP are provided in the following.

6.4.1 THE MAGNETIC MEASUREMENT PROGRAM

MMP was internally developed at CERN in LabVIEW with the main focus on measuring the field in the LHC magnets by means of the rotating coils method. The software includes a control and measurement system. The control system drives the hardware used for the measurements (motors, power supplies, and so on) and monitors the main parameters of the system allowing proper operation to be verified. The user interacts with the system through a graphical user interface. The measurement system measures the field and other parameters to be used in the magnet's analysis.

A principle layout of a typical magnetic field measurements system is shown schematically in Figure 6.2. For each measurement, the software

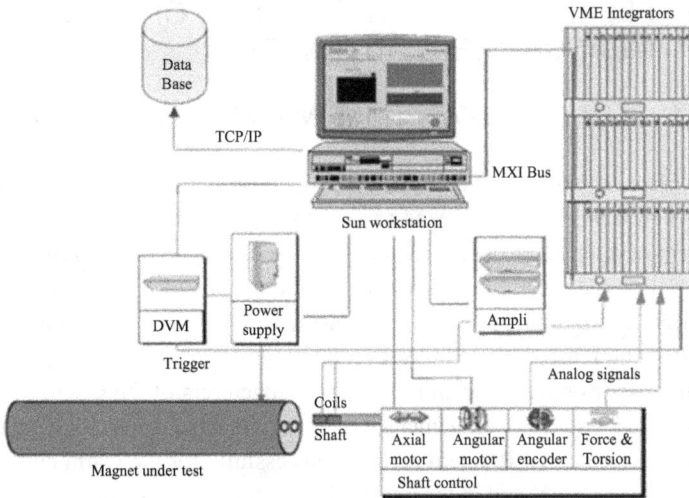

Figure 6.2. Layout of the rotating-coil based measurement system controlled by MMP.

delivers as a result the measured raw data and exploits suitable analysis routines to compute the main field, its direction, the higher-order harmonics, and the magnetic axis coordinates. At the end of each run, the collected raw data are transferred to a database.

The system provides a predefined set of measurement procedures, adjustable to the current needs only through the definition of a limited number of parameters for hardware configuration. As a consequence, the system shows a remarkable lack of flexibility, because it implements a fixed measurement algorithm and an analysis procedure, both based on rotating coils, and requires changes in the LabVIEW code in order to modify them. Each modification therefore requires a long time and the work of expert programmers.

The same approach was used for the development of other programs exploiting different measurement techniques.

As a consequence, before the FFMM project, a plurality of systems was employed, resulting to be not flexible, especially in the test protocol, and difficult to adapt to measurement requirements different from the original design.

This major limitation pushed the research activities, carried out at CERN in the field of magnetic measurements, to move from standalone measurement programs toward the more modern and useful concept of framework.

6.5   FLEXIBILITY REQUIREMENTS FOR MAGNETIC MEASUREMENT AUTOMATION

6.5.1   PAST EXPERIENCES AND NEED FOR FLEXIBILITY

At CERN, the series tests for the LHC's superconducting and resistive magnets were carried out by means of a control and acquisition measurement system developed during the past half a decade. This automatic test system was developed under highly-variable conditions of evolving hardware and software configurations, as well as measurement requirements. As a matter of fact, requirements arose from several test needs incrementally, as well as from preseries requests from different manufacturers of the magnets. The result was a test bench implemented in all major test locations for magnetic measurements at CERN, and adopted successfully for the warm and cold tests both at CERN and in the companies producing the magnets.

The corresponding software for the automation of magnetic measurements at CERN bears a long heritage of the evolution from the original version of the magnetic measurement programs (implemented in C language for a VME bus-based station) to the last version in use in the test stations after the end of the test series campaign for LHC (in excess of 1,000 Sun workstations with Virtual Instruments running on LabVIEW of National Instruments). Data acquisition throughput of this system became to be considered as already too slow during the commissioning period of the LHC with beams, and, therefore, calls for immediate streamlining, as soon as the series tests were completed.

Furthermore, a new hardware of enhanced performance (i.e., digital integrators [21] and rotating units for coils [22]) was developed in order to provide new standards for magnetic measurements. All these conditions demanded a strong redesign and re-engineering of the control and acquisition software as a whole, according to more advanced and unitary paradigms, in order to be adequate for the new measurement requirements and in managing the challenge of the new hardware.

6.5.2   THE PLATFORM FOR MAGNETIC MEASUREMENTS AT CERN

The previous discussion highlights the reasons leading to the launch of the development of a new platform for magnetic measurements at CERN. This new platform was required to evolve from the accumulated knowledge of the developments pursued in the past, by allowing the measurement capability to be extended in harmony with the new available hardware and measurement and test requirements. Although directly aimed at flux measurements (fixed and rotating coils), the new control and acquisition software was expected to bear a large degree of generality to allow extension to other type of measurements (e.g., fast voltage signals from quench detectors in main superconducting magnets, Hall plates, and so on), in order to achieve the benefits of unifying the diverse systems used at that time for superconducting magnets tests.

The aim of the case study presented in this book is to highlight the work carried out to design and prototype a flexible platform for instruments control and data acquisition, integrating the new hardware and software developed suitably, and satisfying a wider range of measurement requirements, variable and evolvable during the time.

6.5.3 HARDWARE OVERVIEW

During the year before the launch of the project for the new platform, a new generation of fast transducers [22, 23] was developed at CERN. These transducers were conceived in order to achieve an increase of up to two orders of magnitude in the bandwidth of harmonic measurements (10 to 100 Hz), when compared to the standard rotating coil technique (typically 1 Hz or less), and still maintaining a typical amplitude resolution of 10 ppm.

In particular, a new *Micro Rotating Unit (μRU)* [22] was designed to turn faster and provide harmonic measurements at rates in the range from 1 to 10 Hz. Fast measurements require that the coils rotate continuously in one direction and at higher speeds (i.e., up to 10 rps). The $μRU$ is capable of turning continuously in one direction up to 8 rps thanks to 54-channel slip rings. The coils are connected in series arbitrarily by means of a patch panel. This permits changes in the compensation schemes or combination of several coils in virtual "super segments," used to measure the integral field.

These developments paved the way for a major improvement of the theoretical and experimental analysis of superconducting accelerator magnets. However, at the same time, they pushed the performance demands on the digital instrumentation used for acquisition [24–26]. Standard magnetic measurements of accelerator magnets require fast and accurate data acquisition with integrating voltmeters. So far, the standard de facto in most subnuclear research centers was the *Portable Digital Integrator* [24]. The core of this instrument is a voltage-to-frequency converter, whose resolution is intrinsically limited by the counting frequency. As a result, this instrument could not follow the evolution of the test requirements arising from the new generation of magnetic transducers described earlier [22], especially considering the increasing need to measure superconducting magnets supplied by high-frequency current cycles and pulses arising at that time [27]. A number of developments worldwide tried to address this issue [25, 26]. At CERN, a multipurpose numerical measurement instrument, the Fast Digital Integrator (FDI), was developed and constituted as one of the main components of the platform hardware [28]. Besides the increased metrological performance [21], the FDI is capable of reducing the flux acquisition time down to 4 μs.

For the development of the FDI, a new generation of high-resolution (18 bit) and high sampling rate (500 kSa/s) ADCs, Successive Approximation Register (SAR) was employed. A DSP was added for on-line processing, thus allowing the decimation of the input samples, with a *Signal-to-Noise Ratio* (SNR) improvement by means of oversampling [21]. After the realization of a first prototype, a digitizer model was developed using the results from experimental tests in order to enhance the design and further improve the performance [29]. Then, the instrument was exploited in the field [21], by achieving (a) ±10 ppm of uncertainty in the measurement of the main field for superconducting magnets characterization, (b) ±0.02% of field uncertainty in quality assessment of small-aperture resistive magnets, and (c) ±0.15% of drift, in an excitation current measurement of 600 s under cryogenic conditions.

6.5.4 SOFTWARE REQUIREMENTS

As far as the software is concerned, the effort for the series test of the LHC superconducting magnets at CERN highlighted limitations in the measurement control and acquisition programs. In particular, main drawbacks arose from the relatively long time needed for the cycle of specification-programming-debugging-validation (i.e., a development iteration). As an example, the aforementioned MMP [19] used at CERN at the time of the LHC series tests had a large spectrum of preprogrammed configurations accessible to the user, but required software specialists for extending the set of configurations to cover new test and analysis requirements. For this reason, more advanced design principles in the field of software engineering became necessary to be considered [16–18].

Furthermore, after the end of the LHC series tests, and during the medium term, the expectation was to have a number of very specific tests to be rapidly adapted and performed on single prototypes or relatively small batches of magnets. This was due to the fact that, apart internal magnet projects for LHC upgrade (e.g., LINAC4 [30]) or study of new accelerators (e.g., CLIC [31]), CERN has been involved mainly in the cooperation for realizing new accelerators over the world (like MedAustron [32], Sesame [33], and so on.) These tests require the control of various devices, such as transducers, actuators, trigger and timing cards, power supplies, and other devices not yet completely specified. Moreover, for different measurement techniques and tests, different algorithms have to be implemented. In practice, the ideal situation would be to have a flexible software framework, providing the tools to help the user in the design of new measurement algorithms, as well as a robust library to control remotely all the instrumentation involved in the tests.

The new system, besides reproducing key operating capabilities of the previous software (reference for comparison is MMP [19]), had to (a) extend the acquisition and control capabilities to the new hardware, and (b) allow user-driven and traceable configuration of the hardware, as well as of the test protocol, in order to bear a maximum capability to evolve.

The new project aimed at maximizing the measurement software quality, in terms of flexibility, reusability, maintainability, and portability, by simultaneously keeping high efficiency levels. In particular, the flexibility, the modification easiness of a system, or component for use in applications or environments, other than those for which it was specifically designed [34], is definitely one of the most desirable properties of any system to face changes in operational environment during its life. This is particularly true for software systems, both, because they are often subject to extremely rapid technological development, and because some of them are specifically conceived to be employed in environments spanning a wide range of functional requirements, not fully predictable at the design stage. This is the case of the new software project, which should be easy to configure for satisfying a large set of measurement applications in the magnetic measurement field.

The main goals of the new project (flexibility, maintainability, reusability, and efficiency) are meant to satisfy the various needs of the different users, according to the classification provided in Table 6.1.

Table 6.1. Main software characteristics and users they address

Software characteristic	User
Flexibility	Test engineer
Maintainability	Developer/administrator user
Reusability	Developer/administrator user, test engineer
Efficiency	Test engineer, end user

6.6 THE FRAMEWORK FFMM

In this section, the software framework realized at CERN (FFMM) is presented by introducing its main design concepts and architecture. Subsequently, its main components, the *Fault Detector* and the *Synchronizer*, are described, by highlighting their architectures and the implementation of their classes. Then, the application of the MDSL (see Section 4.5) and the Advanced User Interface Generator (see Section 4.6) are reported. Finally, implementation details and examples are reported in order to drive the reader to realization.

6.6.1 DESIGN

The FFMM design follows the UML model presented in Chapter 4 (see Figure 4.4). The *TestManager* plays a central role among the device under test, measurands, and the measurement configuration and procedure. The *CommunicationBus* allows the remote control of the *Devices* by *VirtualDevices*. At software level, the measurement timing is performed by the *Synchronizer* (Section 4.4), and the fault and failure detection is managed by *Fault Detector* (Section 4.3). The *Synchronizer* and the *Fault Detector* are, therefore, encapsulated in aspects according to the Aspect-Oriented approach. The scheme of synchronization policy and fault management strategy allows the corresponding modules to be modified, without involving the classes and the framework structure. The layered structure of the framework (Section 4.2.2) is flexible and the functionalities of one layer can be used to form the high level ones. In the following section, the FFMM architecture is presented.

6.6.2 ARCHITECTURE

The framework FFMM has a layered architecture (Chapter 4), shown in Figure 4.2. The internal organization of each layer is object oriented-based, where the objects interact only inside the layer horizontally, among entities of the same level. The upper level capabilities are accessible only through a suitable interface. The architecture presents four different layers: *Basic, Core, Measurement*, and *User* service layers.

In the *Base Service Layer*, subcomponents for environment abstraction, memory management, error handling, file-system abstraction, and processes and threads handling are included. Communication services are defined to allow the extraction of data from devices and external interfaces. In the *Core Service Layer,* measurement devices, event handling infrastructure, fault detection, and logging are included. In the *Measurement Service Layer*, the components for the management of the measurement procedure are defined: (a) the *Test Manager,* (b) the *Measurement Tasks,* and (c) the *Synchronizer* (defined in Section 4.2.2). The *User Service Layer* defines services, such as the User Interface (supported by the GUI engine) and the Integrated Development Environment (supported by the MDSL) for the interaction of the framework with the final users (test engineers and application users).

6.6.3 FAULT DETECTOR

In Figure 6.3, the hierarchy *FaultDetector* of FFMM is reported, by illustrating the static relationships among *Virtual_Device* classes, the

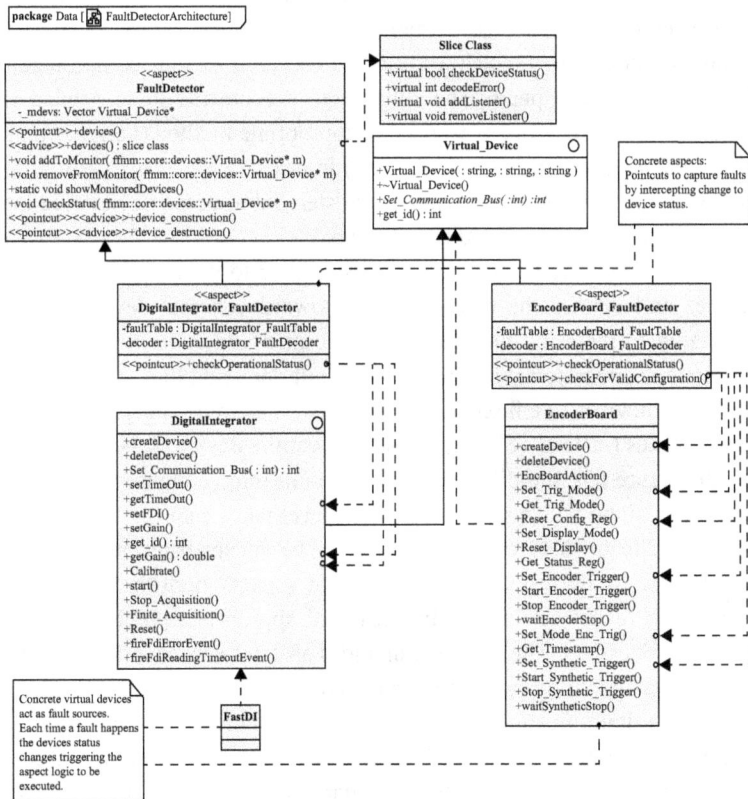

Figure 6.3. An excerpt of the hierarchy of the *FaultDetector*.

FaultDetector aspect with its subaspects, and some concrete virtual devices (*DigitalIntegrator* and *EncoderBoard*). In the figure, the role played by the classes *FaultDecoder* and *FaultTable* is highlighted.

For the sake of clarity, an example related to two devices, typical of magnetic measurement applications, a digital integrator (namely the Fast Digital Integrator, FDI [28]), and an encoder board, with the corresponding classes *FastDI* and *EncoderBoard,* respectively, is reported in the same figure. Encoded fault information is extracted from the *FastDI* device by context interception and is decoded by means of a concrete class *DigitalIntegrator_FaultDecoder*. As an example of AOP design, the aspect *DigitalIntegrator_FaultDetector* is hence responsible for enforcing fault management policies according to the fault kind. It defines the appropriate listener and, when a device is registered in the *FaultDetector*, an instance of the listener is subscribed to the concrete instance of the device. The *FaultDetector* is responsible for defining pointcuts capturing creation and destruction of devices.

The hierarchy *Virtual_Device*, models and organizes all the physical devices involved in the measurement process. Each device has an internal status: modifications to such a status are captured by means of concrete *FaultDetector* subaspects. The subaspects execute the logic needed to decode the modification, as well as an appropriate method (i.e., *fireFault, fireBadParameter, fireError,* and so on) in order to broadcast fault information to the concrete *FaultListener* registered during the device creation and to all the other interested components.

Each *FaultDetector* subaspect is associated to the main devices categories and defines the mapping logic toward concrete device classes belonging to the same family. Indeed, the mapping coarseness between aspects and concrete devices allows the fault detection logic to be reused in similar devices in a flexible manner, as well as to be encapsulated in a few modules (instead of being spread all over the device classes).

In Figure 6.4, the different levels of fault interceptions are depicted according to the fault types. The bottom level takes care of very specific issues and features of concrete devices to be encapsulated in dedicated subaspects. At the middle level, concrete aspects perform continuous monitoring of devices' status, by means of appropriate pointcut expressions and decoders. The top level includes abstract aspects implementing the fault detection logic, reusable in concrete subaspects.

The fault notification strategy has been implemented by means of the architecture "publish-subscribe," such as shown in Figure 6.5, in the case of the device *EncoderBoard*. In particular, the cooperation among *FaultDetector* and its subaspects (in order to associate handlers to fault sources

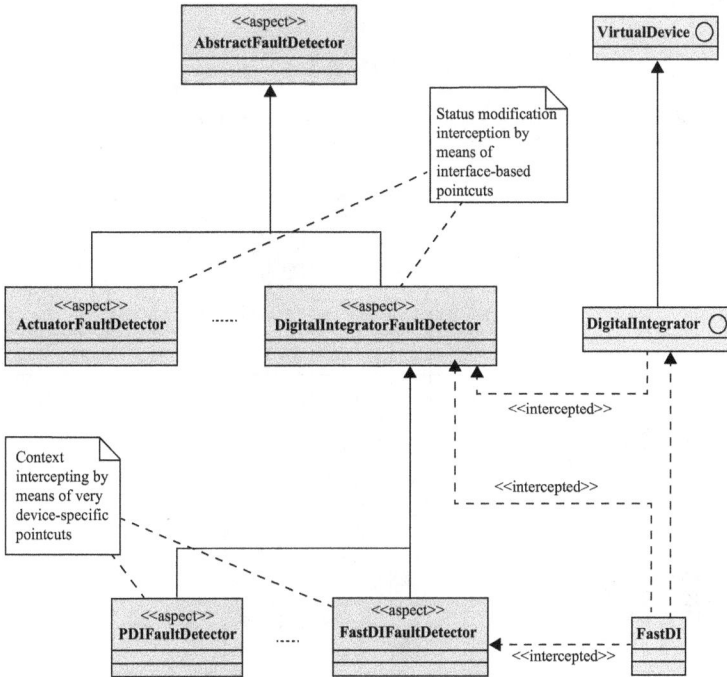

Figure 6.4. Levels of faults interception.

in the measurement system dynamically) is highlighted. The subaspects of the aspect *FaultHandler* have the responsibility to make aware the concrete classes of the faults occurring in the system (e.g., the *TestManger* is responsible for supervising the test session).

The aspect *EncoderBoard_FaultHandler* defines a concrete implementation for the abstract slice class specific for the *EncoderBoard* devices. Such an implementation is responsible for registering and deregistering the *EncoderBoardFaultListener* when an *EncoderBoard* device is created or destructed. Moreover, the subaspect defines the pointcut expressions to intercept and decode faults by calling the appropriate fault broadcasting methods.

This solution allows fault handling logic to be reused in the super-aspects and does not force concrete classes in the system to implement fault handling code. Any component in the system can reply to specific faults occurring anywhere in the system by carrying out the related handling actions.

The concrete classes (*TestManager* or any other components interested in monitoring faults) are oblivious of being faults' handlers, thus the monitoring relationships can be changed by acting simply on aspect

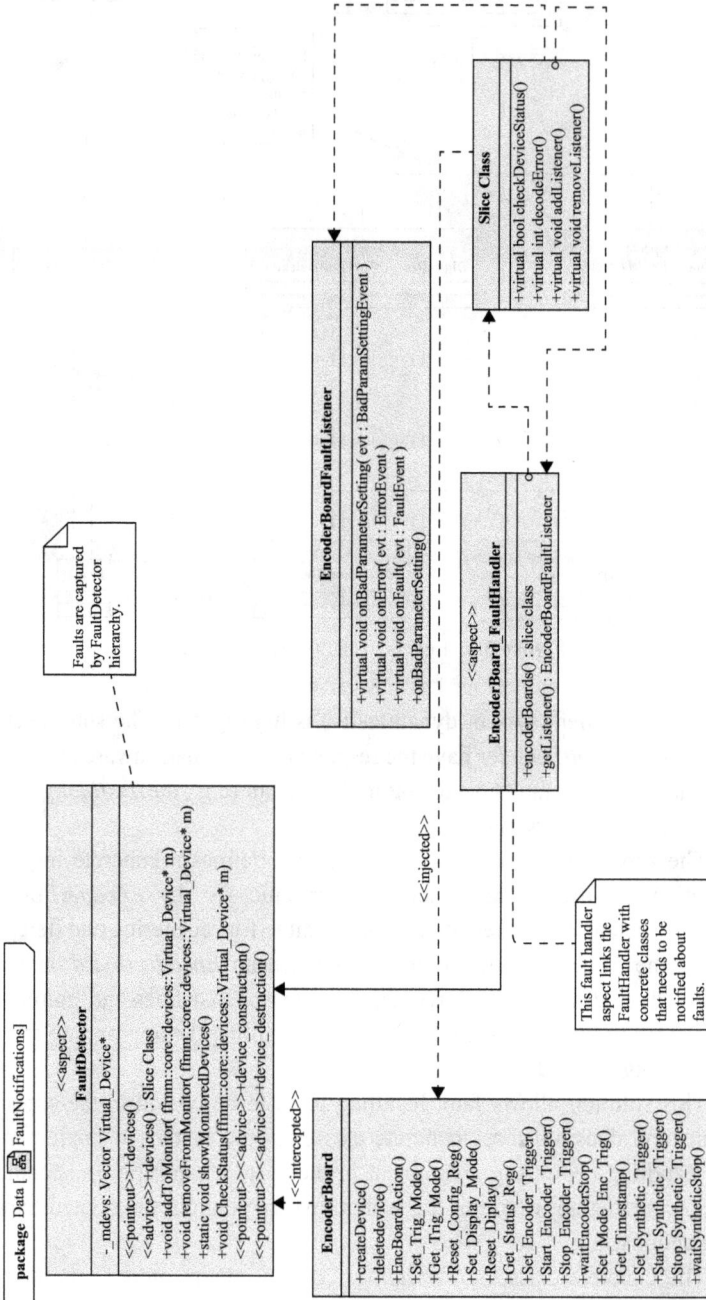

Figure 6.5. Fault notification architecture "publish-subscribe" for the device *EncoderBoard*.

mapping. Shared characteristics among different fault handling logics can be factored out in the aspects, while multiple observations of different kinds of faults can be easily accomplished by defining several listeners for a single concrete class.

In the following, two typical scenarios are discussed in order to highlight the system behavior at run-time. In Figures 6.6 and 6.7, a class and a communication diagram, showing the interception of a device creation and the messages exchanged by key involved components, respectively, are depicted.

In Figure 6.8, a sequence diagram, modeling (a) the typical behavior after listener registration, when faults can happen during normal device operations, and (b) another typical scenario, when a device accesses internal state by means of read and write operations, is reported.

While the proposed architecture is suitable for removing the fault detection code from the devices completely, in a first pilot implementation, the design team decided not to affect the event handling protocol of the framework. Therefore, the code of the fault and error broadcast-

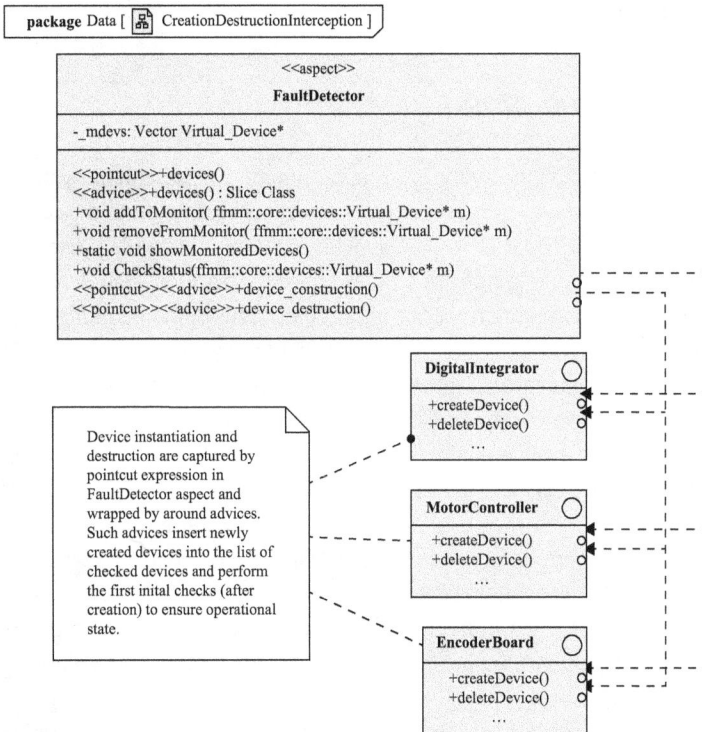

Figure 6.6. *Device Creation/Destruction* interception.

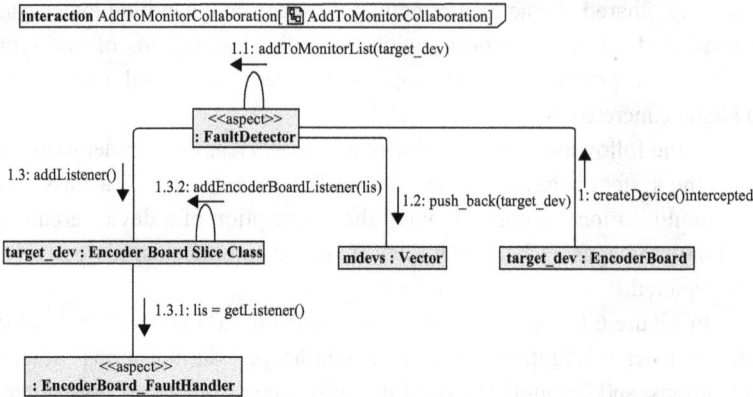

Figure 6.7. Collaboration scenario started by *createDevice* interception on *EncoderBoard*.

ing routines still lies in the component implementing the devices. This approach was aimed at carrying out a concern-driven adoption of AOP.

From this point of view, the event handling is itself a concern to be migrated to the AOP paradigm in future work.

6.6.4 SYNCHRONIZER

As already discussed, a key issue in automatic measurement systems is the capability of assuring proper software synchronization to the test procedure [35–37]. Today, software synchronization is a "widely used technique, and emerging application areas for cost-effective" [38] dependable systems will further increase its importance. The implementation strategies of task synchronization not only affect the system performance, but also its quality, in particular the modularity, reusability, and maintainability [39–42]. Nevertheless, crosscutting concerns can negatively affect the quality of even well modularized synchronization systems implemented by these techniques [43]. In this section, a further example of how AOP [1] can overcome such shortcomings in the implementation of synchronization task is reported. Synchronization tasks are handled more easily by isolating and encapsulating into aspects the crosscutting concerns related to the synchronization of large measurement software. Indeed, in this way only the aspects have to be managed when changes occur, or to reuse them in new systems.

The main goals of the AOP-based design for the software synchronization of measurement tasks are mainly related to achieving better

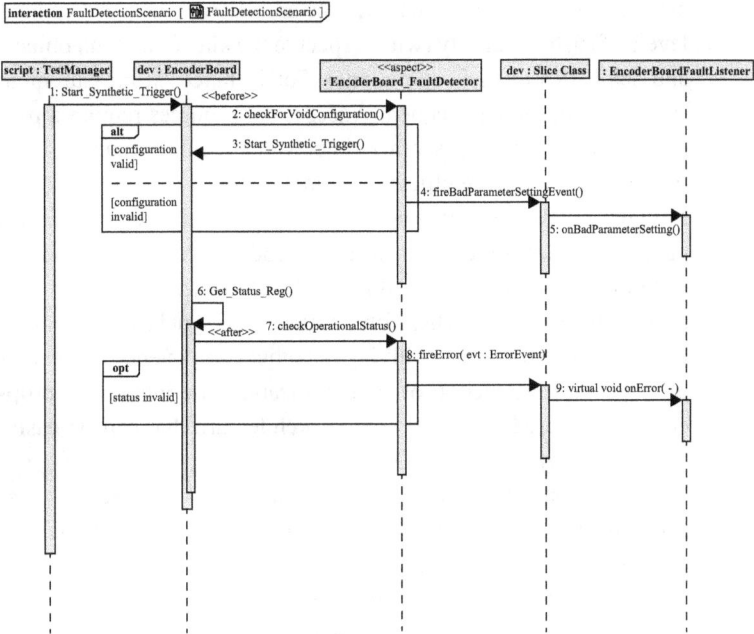

Figure 6.8. Sequence diagram showing the detection of a wrong parameter configuration of an *EncoderBoard* instance.

modularization for concurrency and synchronization concerns while guaranteeing system correctness, better performance, and increased safety.

The solution highlighted here is based on an abstract aspect layer, composed of a simple aspect framework to be reused in the development of synchronization control in different application domains. The modularization achievable through the proposed design makes the synchronization control easy to evolve and simplifies the complexity of the remaining parts of the software, such as devices, fault detection, or logging modules, by decoupling concurrency and synchronization control code from them.

Synchronization is a crosscutting concern particularly hard to modularize through Object-Oriented Programming and design patterns. The AOP-based architecture shown in this book is the result of the analysis of several existing Object-Oriented software systems implementing synchronization and concurrency, which revealed some typical deficiencies of OOP implementation for synchronization. In particular, the lacks are related mainly to the following quality attributes of synchronization software:

•   Extensibility: It is related to the possibility of easily making abstraction of synchronization policies. An optimal synchronization policy cannot be found for all situations, the architecture must assure high levels of customizability (with respect to the kind of domain objects and semantics of their operations). For instance, in measurement systems, objects that frequently access data sources require a pessimistic policy, whereas other objects not requiring such critical timing can use a more efficient optimistic policy.

- Modularity: It is intended as the separation of the code related to the synchronization concern from the code of the base system. In general, synchronization is orthogonal to the other components of a measurement system. Therefore, options like synchronization policy switching (without modifying the other components), as well as incremental introduction of synchronization, are interesting properties considered for the proposed architecture that can be easily enforceable by means of AOP.

- Encapsulation: This is related to the modularity and requires that synchronization is placed within aspects rather than scattered out among synchronized object and their clients. An OOP decentralized design forces to "*spread*" synchronization state in several objects. This means that each component must encapsulate its part of "synchronization" information at run-time even when it doesn't need synchronization (thus wasting memory resources). The AOP architecture enforces, as much as possible, a centralized design in which synchronization state is maintained in the related aspects: components to be synchronized are involved by the aspect encapsulating the synchronization information for all components. When a component doesn't need synchronization no data is stored for it in the aspect and no memory is wasted at all.

- Reusability: Synchronization is a characteristic feature of a measurement system, thus reusability has to be strongly considered and applied to reuse both synchronization and functionality code in an independent fashion. AOP implementation allows aspects implementing synchronization concerns to be reused in an easier way.

6.6.4.1 Architecture

In Figure 6.9, the UML class diagram of the architecture for synchronization in automatic measurement stations, a variant of the design pattern Synchronization Manager based on AOP, is depicted. The stereotype "*aspect*" has been used to distinguish the aspects from the classes defined in the architecture. All the aspects in the diagram are abstract. The

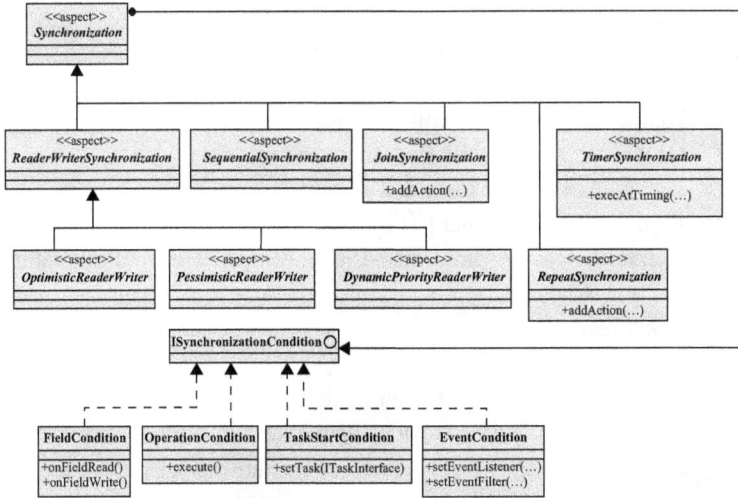

Figure 6.9. AOP-based architecture of a synchronizer for an automatic measurement system.

synchronization aspects can be easily integrated and reused in other architectures and software systems. Indeed, just the components and services to be synchronized, as well as the policy for their synchronization, have to be identified.

The abstract aspect *Synchronizer* provides reusable code and behavior for implementing and modularizing the synchronization logic and policies. Concrete aspects will have the responsibilities of: (a) intercepting components and services interactions to be synchronized and (b) enforcing the right specified policy in the correct context.

The architecture separates the main issues related to the synchronization management: the policy to be adopted, the conditions to be defined, and the specification of the context making use of synchronized elements. The architecture separates these three components allowing the synchronization logic to be reused in the superaspects, without forcing concrete classes in the base system to implement synchronization-handling code. Moreover, concrete classes are oblivious of being synchronized, thus the synchronization policies can be changed by simply acting on aspect mapping.

Common features among different synchronization handling logics can be factored out in the aspects, while multiple observations of different kinds of synchronization can be easily accomplished by defining several mapping aspects for a single concrete class.

In the framework illustrated in this chapter, different aspects are responsible for diverse policies defined in order to support the most interesting scenarios arising from a measurement session. The following policies are defined to synchronize data transfers among devices:

- Joined, repeated, sequential task execution
- Optimistic and pessimistic readers and writers
- Dynamic priority readers and writers
- Producer and consumer policy
- Active devices synchronization supporting different multithreading internal structures (scaling from single thread to a cooperating pool of k threads)

As far as the synchronization conditions are concerned, the architecture provides basic conditions to be aggregated in order to build more complex conditions. They can be used in association with existing or new synchronization policies. The basic implemented conditions are related to field read and write events, operation execution or invocations, and well-defined role execution events.

The resulting architecture is extensible because concrete aspects implementing specific synchronization policies can be easily added and designed to implement new kinds of policies when needed. The added policies only need to implement interface and concrete mapping logic to intercept the client contexts.

The architecture fosters reusability, because existing policies can be reused in several different contexts and the synchronization logic is completely decoupled from the client code. To summarize the benefits of an Aspect-Oriented approach to software synchronization in automatic measurement systems key advantages are just code maintainability and reusability. For each new device added to the station, the related synchronization code can be added to the synchronization hierarchy. All the code is encapsulated in few subaspects, thus common features among different synchronization logics are well structured and factored out. As a consequence, the AOP Synchronizer design, with respect to a traditional OOP version, exhibits a better modularized design by eliminating code scattering and tangling, and increasing the possibility of code reuse.

6.6.5 MEASUREMENT DOMAIN SPECIFIC LANGUAGE

The implementation of the Measurement Domain Language (MDSL) inside FFMM exploits main features of the Eclipse Modeling Project [44, 45], a modeling framework and code generation facility for building tools and other applications. Eclipse is based on a structured data model consisting of

- interfaces and implementation classes for all the classes in the model, plus a factory and package (meta data) implementation class;
- implementation classes that adapt the model classes for editing and display; and
- a properly structured editor, conforms to the recommended style for model editors of Eclipse Modeling Framework (EMF), serving as a starting point for customization.

In FFMM, the MDSL is conceived as a form of user interface [46] aimed at providing an additional view of FFMM, specifically conceived for the test engineer. The developer can operate with C++ at any level in the system, including the definition of a measurement script. On the other hand, the test engineer, with limited effort and programming skills, can operate at script level by means of the DSL. In Figure 6.10, a schematic representation of the MDSL implementation in FFMM is provided. The Measurement Domain Specific Description (MDSD) contains the Measurement Test Procedure written by the test engineer.

First, this script is read by the MDSL Parser, subsequently the Builder generates the C++ code performing the required test procedure by means of FFMM classes.

In particular, MDSL has been developed by the Eclipse's features of *Xtext* and *Xpand* [45]. *Xtext* is a tool for developing domain-specific and textual programming languages. It provides a language-specific integrated

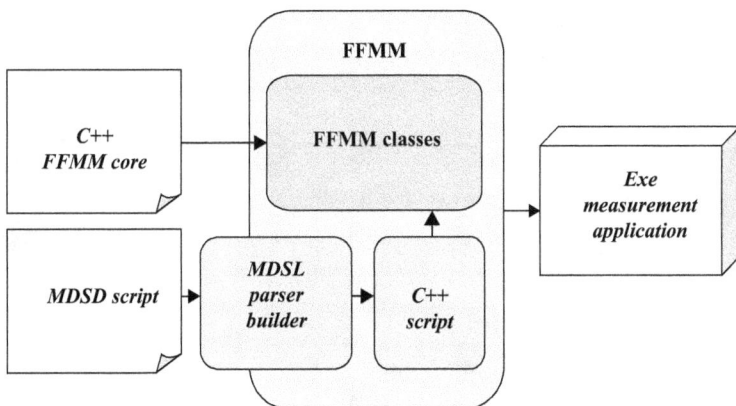

Figure 6.10. MDSL implementation in FFMM.

development environment and, in addition to common parser generators (like e.g., JavaCC [47] or ANTLR [48]), generates

- an incremental, ANTLR3-based parser to read models from MDSD script;
- a serializer, both to write models back to text and to produce the measurement specific semantic model;
- a linker, to establish cross links between model elements;
- the integration of the language into the Eclipse integrated development environment (IDE).

In *Xtext,* the script structure can be set up by embedding the rules of the new language specific grammar. In Figure 6.11, a resumed example of MDSL grammar implementation is shown.

```
Script:
'BEGIN_SCRIPT' scriptName = STRING ':'
        (meastasks += MeasurementTask)*
'END_SCRIPT';


MeasurementTask:
        'BEGIN_MTASK' taskName = STRING ':'
                (DeviceDefinitions += DefinitionStatement)*
                (DeviceConfigurations += ConfigurationStatement)*
                (DeviceCommands += CommandStatement)*
        'END_MTASK';


DefinitionStatement:
        'DEF' (
                motor_defs = Def_Motor_
        |       fdi_defs = Def_Fdi_ ) ';';


ConfigurationStatement:
        'CFG' (
                motor_cfgs = Cfg_Motor_
        |       fdi_cfgs =  Cfg_Fdi_ ) ';';


CommandStatement:
        'CMD' (
                motor_cmds = Cmd_Motor_
        |       fdi_cmds =  Cmd_Fdi_ ) ';';
```

Figure 6.11. Specific grammar example of MDSL in *Xtext.*

According to the example: a script (between the hot words *BEGIN_ SCRIPT* and *END_SCRIPT*) can contain only measurement task statements; a task (between the hot words *BEGIN_MTASK* and *END_MTASK*) can be composed of either device definition statements (starting with *DEF*), or device configuration statements (starting with *CFG*), or device command statements (starting with *CMD*), and the order must be respected. In the statements declaration organized by typology, for example, the definition statement in Figure 6.11, the list of permitted methods is specified. If the item is preceded by the symbol "|", the order is not mandatory. Basically, a grammar tree can be defined in *Xtext* by adding the grammar rules from the generic to the specific ones. In particular, the example is tailored to the FDI and a motor controller, specific devices exploited in the Case Studies Section.

Some of the IDE features, either derived from the grammar or easily implementable, are: syntax coloring, model navigation, code completion, outline view, and code templates.

Xpand is a tool specializing in code generation based on EMF models and used as a builder to produce the C++ code suitable for FFMM framework. Each object already defined in the grammar (*Xtext*), has to be treated by the Domain-Specific Builder (*Xpand*). The rules of code generation are specified in the *Xpand* file. In Figure 6.12, two examples of builder rules, for the FDI instrument and a motor controller, are pointed out.

By reiterating the same process for all the devices supported by FFMM, the whole MDSL has been developed. In general, all the commands related to standard acquisition boards, multimeters, digital integrators, precision positioning stages, motor controllers, and encoder boards are already implemented. In addition, some commands related to user interface programming and the data presentation are included [49]. To

```
// Fdi configuration method
«DEFINE Fdi_ConfigurationMethods FOR Cfg_Fdi_ -»
«nameFdiDefined»->configure(«int_bus-», «int_slot-» );
«ENDDEFINE»


// Motor controller configuration method
«DEFINE motor_ConfigurationMethods FOR Cfg_motor_ -»
«namemotorDefined»->configure(«int_serial_port-»);
«ENDDEFINE»
```

Figure 6.12. Domain-specific Builder rules of MDSL in *Xpand*.

give an idea, the effort for adding a new device into the specific grammar of the already defined language can be quantified in about 20 new lines code for the device creation and about 10 lines for each method by which the device is composed.

6.6.6 ADVANCED USER INTERFACE GENERATOR

The Advanced User Interface Generator is an important component of the framework, because it allows the test engineer to easily produce an Application User Interface permitting the final user (e.g., the test technician) to interact with the software application. In a script-based application in FFMM, all the software functions should be set and configured by the script. This chapter highlights (a) the configuration and the use of the graphical user interface within the script, (b) how to perform the input and output of values by using the Graphic Interaction Component, and (c) how to use the plotting features.

6.6.6.1 Configuration of the Graphical User Interface in the User Script

In the FFMM measurement script, the test engineer can select which graphical window he wants to show during the measurement procedure, by defining their appearance order and their effect on the script. At the beginning of each script, the main window (Figure 6.13), with the basic controls, can be activated by adding the macro **MAIN_WINDOW.**

In the main window, there is:

- a file and directory selector, where the user will insert the name and the path of all the result files of the measurement;
- the main script controls, in particular the buttons:
  - *start*, to trigger the start of the script execution;
  - *stop*, to stop the measurement (only if needed);
  - *abort*, to launch a fatal fault from the user;
  - *exit*, to force the end of the program;
- a log area, where the log information are displayed.

Just like the main window, other windows can be activated from the script. For example, to obtain the user's configuration parameters needed to set up a FDI Cluster, the test engineer can use the *FdiClusterFrame*

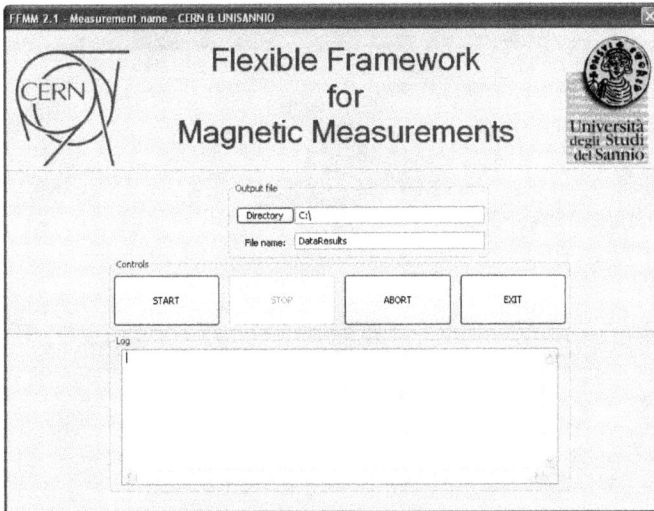

Figure 6.13. Main window.

```
FdiClusterFrame* ffdi = new FdiClusterFrame();

ffdi->ShowModal();

dataFDIcluster* fdi_cl = (dataFDIcluster*)
Parameters::get("FDI",1);


int* bus = fdi_cl->_bus;

int* slot = fdi_cl->_slot;

int number_of_FDI = fdi_cl->_num_FDI;

double* gain = fdi_cl->_gain;
```

Figure 6.14. Code for the *FdiClusterFrame*.

(Figure 6.14). The next code fragment shows how the window can be created and displayed and how the data can be retrieved from it.

In Figure 6.15, the window look is shown.

In the same way, the *EncoderFrame*, a window asking the application's user the parameters for the encoder board setting, can be displayed (Figures 6.16 and 6.17).

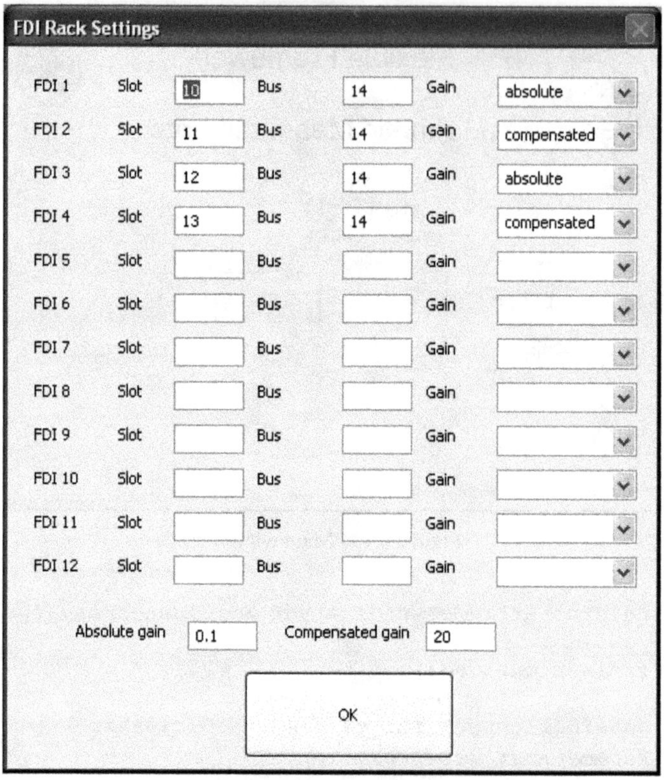

Figure 6.15. *FdiClusterFrame* window.

```
EncoderFrame* fenc = new EncoderFrame();

fenc->ShowModal();

dataENC* enc = (dataENC*) Parameters::get("ENC", 1);


int bus = enc->m_bus;

int slot = enc->m_slot;

int number_point_per_turn = enc->m_num_ppt;;

char division_ratio = enc->div_r;

int rotation = enc->rot;
```

Figure 6.16. Code for the *EncoderFrame*.

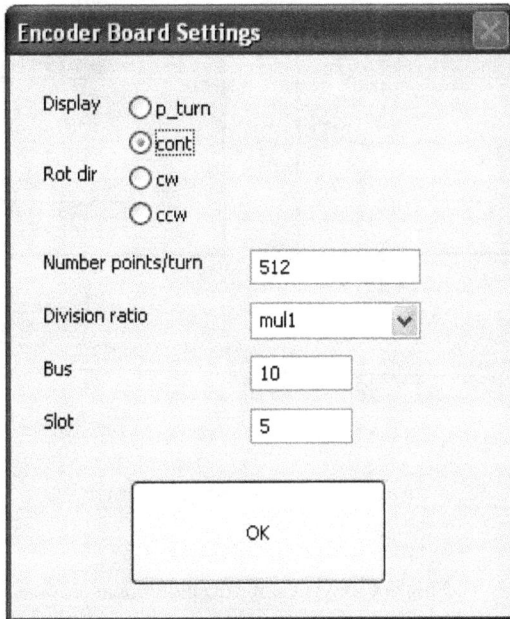

Figure 6.17. *EncoderFrame* window.

```
GIC<int> i_gic("i");

int i[3];

i_gic.input(i,3,"Input 3 int values:","value ");
```

Figure 6.18. Code example for the *GIC*.

6.6.6.2 Input/Output of Values with the Graphic Interaction Component

During a measurement procedure, information is exchanged several times between the user and application; the test engineer can decide when and where in the script the input and output (I/O) process takes place and the values (how many and which type) are involved. Generally, for the I/O in the configuration or setting phase of the script, the *Graphic Interaction Component* (*GIC*) can be used. The *GIC* class is a template: When declared, as shown in Figure 6.18, the type (char, int, float, double, and so on) is preset and the object is created in order to manage the values of

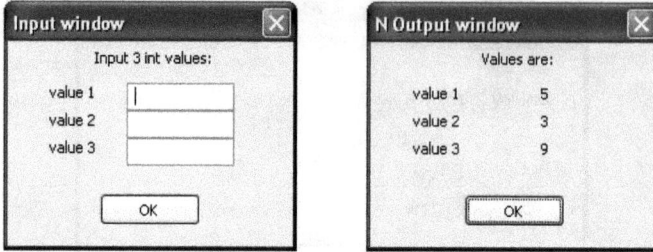

Figure 6.19. Windows generated by the *GIC*.

```
double data[4096];

FPlot* fp1 = new FPlot();

fp1->PlotArray(data, 1024);

fp1->Xlabel("X label");
```

Figure 6.20. Code for the plot function.

selected type. In the figure, the code fragment shows how to instantiate a GIC object of the integer type and how to use its functionalities.

The *GIC* class provides two methods for the interactive I/O:

- *input(...)*, to ask the user some values
- *output(...)*, to display on the screen some values

In Figures 6.19, the resulting windows in the case of an integer *GIC* are shown.

6.6.6.3 *Plotting Functions of Numeric Values*

Another important feature of the graphic interface is the plotting functions. If during the measurement procedure a set of value, for example, an array of double, has to be displayed in a plot format, the graphic functionalities of the *FPlot* can be used as shown in Figure 6.20.

If data come out from an acquisition device, such as an FDI, there is a proper way to plot them on line. As shown in Figure 6.21, the class *FPlot* has a useful method (*SetFdi(...)*) to connect the plot (Figure 6.22) with the data coming from the device and to display a window auto-updating with the rate specified.

```
FastDI* fdi;

fdi = FastDI::createDevice("fdi1");

int plotRate = 10;

FPlot* fp2 = new FPlot();
```

Figure 6.21. Code for the FDI data plot function.

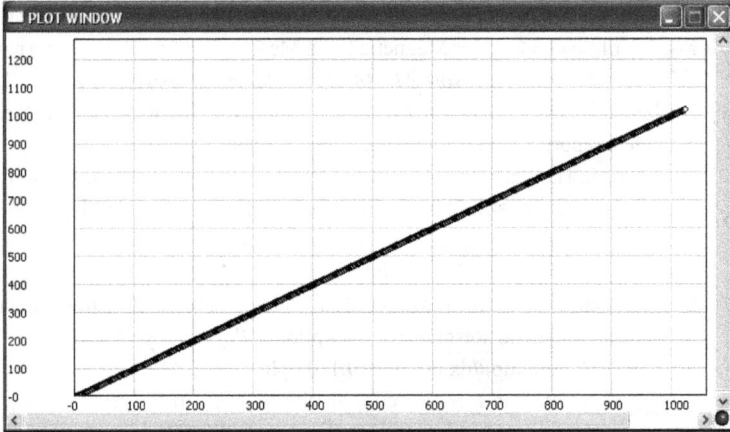

Figure 6.22. Plot window.

REFERENCES

[1] Arpaia, P., M. Buzio, L. Fiscarelli, and V. Inglese. November 2012. "A Software Framework for Developing Measurement Applications Under Variable Requirements." *AIP Review of Scientific Instruments* 83, no. 11. doi: 10.1063/1.4764664

[2] Blundell, S.J. 2009. *Superconductivity: A Very Short Introduction.* New York, NY: Oxford. ISBN-10: 019954090X.

[3] Gareyte, J. 1996. "Impact of superconductors on LHC design." In *CERN 96-03*, pp. 335–46, Geneva, Switzerland: CERN.

[4] CERN Communication Group. February 2009. *CERN LHC Guide*, https://cds.cern.ch/record/1165534/files/CERN-Brochure-2009-003-Eng.pdf

[5] Elmore, W.C., and M.W. Garrett. 1954 "Measurement of Two-dimensional Fields, Part I: Theory." *Review of Scientific Instrument* 25, no. 5, pp. 480–5. doi: http://dx.doi.org/10.1063/1.1771105

[6] Dayton, I.E., F.C. Shoemaker, and R.F. Mozley. 1954. "Measurement of Two-dimensional Fields, Part II: Study of a Quadrupole Magnet." *Review of Scientific Instruments* 25, no. 5, pp. 485–9. doi: http://dx.doi.org/10.1063/1.1771107

[7] Jain, A.K. April 1997. "Harmonic Coils." CERN Accelerator School Proceedings, pp. 175–217. Anacapri, Italy: CERN Document Server.

[8] Bottura, L., and K.N. Henrichsen. 1997. "Standard Analysis Procedures for Field Quality Measurement of the LHC Magnets—Part I: Harmonics." *CERN Internal note EDMS 313621*. Geneva, Switzerland: CERN Document Server.

[9] Bottura, L., and K.N. Henrichsen. May 2002. "Field Measurements." *CERN Accelerator School Proceedings*. Erice, Sicily, Italy: CERN Document Server.

[10] Bidon. S., J. Billan, F. Fischer, and C. Sanz. 1995. "New Technique of Fabrication of Search Coil for Magnetic Field Measurement by Harmonic Analysis." In *CERN Internal Note AT-MA 95-117*. Geneva, Switzerland: CERN.

[11] DiMarco, J., and J. Krzywinsky. March 1996. Mtf Single Stretched Wire. Technical report, Fermi National Accelerator Laboratory.

[12] DiMarco, J., H. Glass, M.J. Lamm, P. Schlabach, C. Sylvester, J. C. Tompkins, and J. Krzywinsky. 2000. "Field Alignment in Quadrupole Magnets for the LHC Interaction Region." *IEEE Transactions on Applied Superconductivity* 10, No. 1, pp. 127–30. doi: http://dx.doi.org/10.1109/77.828192

[13] Bloenbergen, N., Purcell, E.M., and R.V. Pound. 1948. "Relaxation Effects in Nuclear Magnetic Resonance Absorption." *Physical Review* 73, no. 7, pp. 679–712. doi: http://dx.doi.org/10.1103/physrev.73.679

[14] Hall, E.H. 1879. "On a New Action of the Magnet on Electric Currents." *American Journal of Mathematics* 2, no. 3, pp. 287–92. doi: http://dx.doi.org/10.2307/2369245

[15] Pearson, G.L. 1948. "A Magnetic Field Strength Meter Employing the Hall Effect in Germanium." *Review of Scientific Instruments* 19, no. 4, pp. 263–65. doi: http://dx.doi.org/10.1063/1.1741240

[16] Guerrero, A., J-J. Gras, J-L. Nougaret, M. Ludwig, M. Arruat, and S. Jackson. 2003. "CERN Front-end Software Architecture for Accelerator Controls." In *Proceedings of ICALEPCS2003*. Gyeongiu, Korea: CERN.

[17] Nogiec, J.M., J. Di Marco, S. Kotelnikov, K. Trombly-Freytag, D. Walbridge, and M. Tartaglia. Jun. 2006. "Configurable Component-based Software System for Magnetic Field Measurements." *IEEE Tranactions On Applied Superconductivity* 16, no. 2, pp. 1382–85. doi: http://dx.doi.org/10.1109/tasc.2005.869672

[18] Abeille, G., S. Pierre-Joseph, J. Guyot, M. Ounsy, S. Rubio, G. Strangolino, and R. Passuello. October 10–14, 2011. "TANGO Archiving Service Status." *13th International Conference on Accelerator and Large Experimental Physics Control Systems*, pp. 127–30. Grenoble, France: Joint Accelerator Conferences Website (JACoW).

[19] Madaro, L., A. Rijllard, R. Saban, L. Walckiers, L. Bottura, and P. Legrand. 1996. "A VME-Based Labview System for the Magnetic Measurements of the LHC Prototype Dipoles." In *Proceedings of EPAC 96*. Barcelona, Spain: CRC Press.

[20] Walckiers, L. May 1992. "The Harmonic Coil Method." In *CERN Accelerator School on Magnetic Measurements and Alignment.* Geneva, Switzerland: CERN.

[21] Arpaia, P., L. Bottura, L. Fiscarelli, and L. Walckiers. February 2012. "Performance of a Fast Digital Integrator in On-field Magnetic Measurements for Particle Accelerators." *AIP Review of Scientific Instruments* 83, no. 2. doi: 10.1063/1.3673000

[22] Brooks, N.R., L. Bottura, J.G. Perez, O. Dunkel, and L. Walckiers. June 2008. "Estimation of Mechanical Vibrations of the LHC Fast Magnetic Measurement System." *IEEE Transactions on Applied Superconductivity* 18, no. 2, pp. 1617–20. doi: http://dx.doi.org/10.1109/tasc.2008.921296

[23] Haverkamp, M., L. Bottura, E. Benedico, S. Sanfilippo, B. ten Haken, and H.H.J. ten Kate. March 2002. "Field Decay and Snapback Measurements Using a Fast Hall Plate Detector." *IEEE Transactions on Applied Superconductivity* 12, no. 1, pp. 86–89. http://dx.doi.org/10.1109/tasc.2002.1018357

[24] Galbraith, P. 1993. "Portable Digital Integrator." In *Internal Technical Note 93-50, AT- MA/PF/fm.* Geneva, Switzerland: CERN.

[25] Evesque, C. September 1999. "A New Challenge in Magnet Axis Transfer." In *Proceedings of International Magnetic Measurement Workshop IMMW11.* Upton, NY: Brookhaven National Laboratory (USA).

[26] Carcagno, R., J. DiMarco, S. Kotelnikov, M. Lamm, A. Makulski, V. Maroussov, R. Nehring, J. Nogiec, D. Orris, O. Poukhov, F. Prakoshin, P. Schlabach, J.C. Tompkins, and G.V. Velev. 2006. "A Fast Continuous Magnetic Field Measurement System Based on Digital Signal Processor." *Applied Superconductivity, IEEE Transactions* 16, no. 2, pp. 1374–77. http://dx.doi.org/10.1109/TASC.2005.869702

[27] Pellico, W., and P. Colestock. May 12–16, 1997. "Pulsed Magnetic Field Measurement Using a Ferrite Waveguide in a Phase Bridge Circuit." In *Proceedings of the Particle Accelerator Conference* 3, pp. 3716–8. Vancouver, BC: IEEE.

[28] Arpaia, P., A. Masi, and G. Spiezia. April 2007. "A Digital Integrator for Fast Accurate Measurement of Magnetic Flux by Rotating Coils." *IEEE Transactions on Instrumentation and Measurement* 56, no. 2, pp. 216–20. doi: 10.1109/TIM.2007.890787

[29] Arpaia, P., V. Inglese, G. Spiezia, and S. Tiso. June 2009. "Surface-Response-Based Modeling of Digitizers: A Case Study on a Fast Digital Integrator at CERN." *IEEE Transactions on Instrumentation and Measurement* 58, no. 6, pp. 1919–28. doi: http://dx.doi.org/10.1109/tim.2008.2005855

[30] Arpaia, P., M. Buzio, O. Dunkel, D. Giloteaux, and G. Golluccio. May 3–6, 2010. "A Measurement System for Fast-Pulsed Magnets: A Case Study on Linac4 at CERN." *IEEE International Instrumentation and Measurement Technology Conference.* Austin, TX: IEEE.

[31] Battaglia, M., A. De Roeck, J. Ellis, and D. Schulte. 2004. Physics at the CLIC Multi-TeV Linear Collider. Technical Report, Report of the

CLIC Physics Working Group, CERN Report, ref. hep-ph/0412251, CERN-2004-005.

[32] Golluccio, G., et al. June 3–7, 2013. "Overview of the Magnetic Measurements Status for the MedAustron Project." *18th International Magnetic Measurement Workshop (IMMW18)*, Brookhaven, GA. Online: https://indico.bnl.gov/conferenceDisplay.py?confId=609

[33] Sesame Computing Group. 2011. SESAME—Synchrotron-light for Experimental Science and Applications in the Middle East, http://www.sesame.org.jo/sesame/

[34] IEEE. 1999. *Standard Glossary of Software Engineering Terminology 610.12-1990*, Vol. 1. Los Alamitos, CA: IEEE Press.

[35] Gupta, R.K., C.N. Coelho, and G. De Micheli. 1992. "Synthesis and Simulation of Digital Systems Containing Interacting Hardware and Software Components." In *Proceedings of the 29th ACM/IEEE Design Automation Conference*, pp. 225–30. Anaheim, CA: IEEE.

[36] Graunke, G., and S. Thakkar. June 1990. "Synchronization Algorithms for Shared Memory Multiprocessors." *Computer* 23, no. 6, pp. 68–69. doi: http://dx.doi.org/10.1109/2.55501

[37] Praum, C. von, H.W. Cain, J. Choi, and K.D. Ryu. 2006. "Conditional Memory Ordering." In *Proceedings of the 33rd IEEE ISCA*, pp. 41–52. Yorktown Heights, NY: IBM T.J. Watson Research Center.

[38] Arpaia, P., L. Fiscarelli, G. La Commara, and F. Romano. January 2011 "A Petri Net-Based Software Synchronizer for Automatic Measurement Systems." *IEEE Transactions on Instrumentation and Measurement* 60, no. 1, pp. 319–28. doi: http://dx.doi.org/10.1109/tim.2010.2046602

[39] Bosch, S.J. 1999. "Design of an Object-Oriented Framework for Measurement Systems." In *Domain-specific Application Frameworks*, eds. M. Fayad, D. Schmidt, and R. Johnson, pp. 177–205. New York, NY: John Wiley ISBN: 0-471-33280-1.

[40] Wang, S., and K.G. Shin. August 2002. "Constructing Reconfigurable Software for Machine Control Systems." *IEEE Transaction on Robotics and Automation* 18, no. 4, pp. 474–86. doi: http://dx.doi.org/10.1109/tra.2002.802235

[41] Beck, J.E., J.M. Reagin, T.E. Sweeney, R.L. Anderson, and T.D. Garner. June 2000. "Applying a Component-based Software Architecture to Robotic Workcell Applications." *IEEE Transaction on Robotics and Automation* 16, no. 3, pp. 207–17. doi: http://dx.doi.org/10.1109/70.850639.

[42] Jennings, N.R. 1999. "Agent-based Computing: Promises and Perils." In *Proceeding of the sixteenth International Joint Conference on Artificial Intelligence (IJCAI)*, vol. 2, pp. 1429–36. Stockholm, Sweden: Morgan Kaufmann.

[43] Pfister, C., and C. Szyperski. July 1996. "Why Objects Are Not Enough." In *Proceeding of the First International Component Users Conference (CUC)*. Munich, Germany: SIGS Publishers.

[44] Steinberg, D., F. Budinsky, M. Paternostro, and E. Merks. December 2008. *EMF: Eclipse Modeling Framework.* 2nd ed. Reading, MA: Addison-Wesley.

[45] Gronback, R.C. March 2009. *Eclipse Modeling Project: A Domain-Specific Language (DSL) Toolkit.* 1st ed. Reading, MA: Addison-Wesley.

[46] Bosch, J., and G. Hedin. April 1996. "Editors's Introduction." In *Proceedings ALEL'96 Workshop on Compiler Techniques for Application Domain Languages and Extensible Language Models, Technical Report LU-CS-TR,* pp. 96–173. Lund, Sweden: Lund University.

[47] JavaCC is a Parser/Scanner Generator. January 2011. *Java.net,* http://java.net/projects/javacc/

[48] Parr, T. December 2009. Language Implementation Patterns: Create Your Own Domain-Specific and General Programming Languages. 1st ed. Raleigh, NC: Pragmatic Bookshelf.

[49] Arpaia, P., L. Fiscarelli, and G. La Commara. February 2010. "Advanced User Interface Generation in the Software Framework for Magnetic Measurements at CERN." *Metrology and Measurement Systems* 17, no. 1, pp. 27–38. doi: http://dx.doi.org/10.2478/v10178-010-0003-y

CHAPTER 7

IMPLEMENTATION

It is not enough for code to work.

—Robert C. Martin, *Clean Code:
A Handbook of Agile Software Craftsmanship
tags: software 7 likes like*

Every successful hardware has a software behind

—Thiru Voonna

7.1   OVERVIEW

In this chapter, a few implementation examples in C++ code of the most significant parts of the Flexible Framework for Magnetic Measurements (FFMM) at CERN are illustrated. The examples are chosen by referring to the main layers of the framework architecture. In particular, for the base service layer, implementation details for some of the active devices and transducer classes are shown. For the core service layer, coding particulars both about *FaultDetector*, the main class implementing the service of fault detection, and *IFault*, the interface providing the prototypes of the methods used by the service are presented. For the measurement service layer, the implementation of both the Petri net's based component aimed at managing the software synchronization and its main classes, *Labeled Petri Net* and *Synchronizer*, are shown in detail. For the user service layer, in general, the creation of a new project in Eclipse, in order to define a new domain specific language and, in particular, the implementation of the Measurement Domain Specific Language (MDSL) are illustrated. Finally, the chapter is completed by the analysis of the software quality of the FFMM source code according to the standard ISO 9126: In particular, main code quality metrics are computed by the tool Understand C++ and architecture and design quality are assessed through the overview pyramid by the tool inFusion.

7.2 BASE SERVICE LAYER

In the following, implementation examples of the classes, (a) *CommunicationBus*, (b) *Active Devices*, and (c) *Transducer* of the Base Service Layer are detailed.

7.2.1 *CLASS* COMMUNICATIONBUS

An automatic measurement system is composed of several cooperating instruments. In most cases, the devices are connected to a central control node (normally a PC) by means of different communication buses (e.g., IEEE488, PXI), handled by software components (communication services) grouped in the base service layer of the framework architecture.

The communication services are designed according to the pattern "Service Configurator" [1], for the following main motivations:

- A communication service must be initiated, suspended, resumed, and terminated dynamically.
- A communication between the device and the controller is simplified by exploiting multiple independently developed and dynamically configurable communication services.
- Multiple communication services can be managed easily or optimized by configuring them through a single administrative unit.

Therefore, the configurator *CommunicationBus* [2] was introduced with the following key participants (Figure 7.1):

- *ICommunicationBus*, specifying the interface containing the abstract methods (e.g., methods for initialization and termination) used

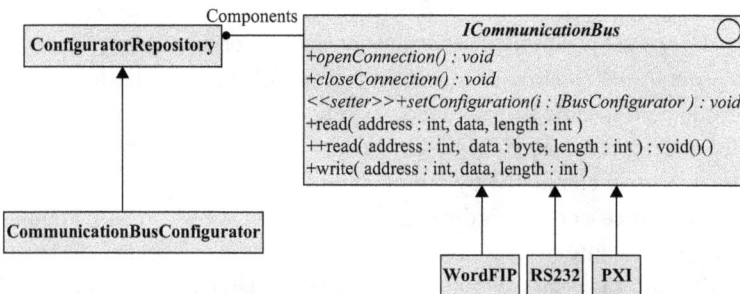

Figure 7.1. Structure of the configurator *CommunicationBus*.

by a *CommunicationBusConfigurator*-based application to dynamically configure each concrete communication bus.

- Concrete communication bus (e.g., *WordFIP. RS232, PXI*), implementing the *ICommunicationBus* methods.
- *ConfiguratorRepository*, maintaining a repository of all communication buses. This allows the behavior of the communication services to be managed and controlled by an administrative entity.

The behavior of the class managing the communication services *CommunicationBus* is characterized by three main phases (not all mandatory):

- *Bus initialization.* The configurator dynamically creates, initializes, and adds a component to its repository that manages it at run-time.
- *Bus use.* In an application, a communication device performs its processing tasks. The component configurator can suspend and resume existing communication buses temporarily, for example, during a reconfiguration.
- *Bus termination.* The configurator has the responsibility of shutting down communication buses when they are no longer needed, in order, to clean up their resources before terminating. After termination, the configurator removes the communication bus from the component repository and unlinks it from the application's address space.

Such a design is capable of providing the communication with an efficient centralized administration. Furthermore, modularity and reusability are improved. A communication bus does not need a predefined configuration at run time, but it could be configured according to user-specified settings. Figure 7.2 shows the class diagram of the communication services architecture. The interface *IBusConfigurator* includes a method for identifying the concrete bus configurator.

For each physical bus, a concrete configurator inherits *IBusConfigurator* and implements a specific behavior. If a device changes its communication channel (i.e., from RS232 to USB), only the concrete bus configurator is to be substituted with another appropriate type.

The interface *ICommunicationBus* defines a mechanism for sending and receiving data to and from components in an abstract way. Specific implementations are defined in the concrete communication buses (e.g., the class *PXI*). The configuration at run time is carried out via a set method in the communication bus.

A concrete *CommunicationBus* is instantiated by the static method *CreateCommunicationBus*, provided by the class *FactoryBus*. Each device uses the *FactoryBus* to obtain the specific bus instance depending

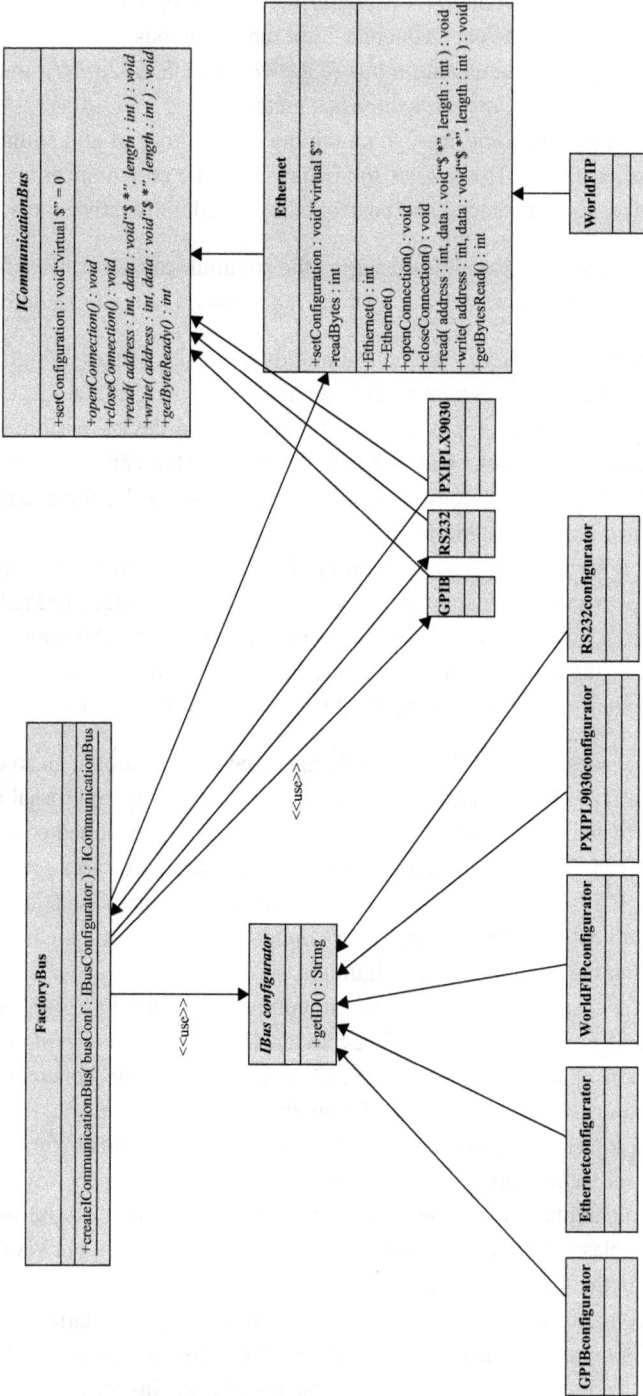

Figure 7.2. Structure of the communication services.

on the configurator parameters. Bus implementation details are hidden to devices, only the configurator has to be known.[1]

7.2.2 CLASS ACTIVE DEVICE

The class *Active device* [3] allows a wide reusability and flexibility of the FFMM infrastructure to be achieved. The main design concept is the view of objects as concurrent processes according to the design pattern "devices like active objects" [3]. The "active object" can change its internal state according to its internal operations executing within the object itself. The object's process recurrently executes these operations without any suspension due to other running processes.

Conceptually, an object or device requests a service from another object, by sending a "service request" message. Apart from specific cases to be treated separately, a serial execution, where an object waits for the completion of a requested service before starting another one, is not necessarily needed. Thus, the general model of interaction allows objects to be executed concurrently as parallel processes [4].

In this context, an efficient strategy of measurement synchronization can be introduced, where

- an *active* entity communicates with other objects by sending messages;
- an object is composed of data, a communication module, an incoming message queue, and its own execution context;
- each active object communicates only by sending messages, thus they can run concurrently (each device runs in a separate thread).

A first design of this "active object" concept was applied to the logging functions of FFMM [5]. Figure 7.3 highlights the communication infrastructure between the active "object devices" and the *Logging subsystem*, used to store the data acquired by the measurement devices.

The *FFMMLogger*, the *Synchronizer* and the *Devices* can be viewed as concurrent objects running in separate threads. As soon as the user (or the *TestManager* [6]) asks for saving data, the *Synchronizer* sends a *trigger message* to the devices involved in the measurement session and to the *Logger Receiver*. When the trigger message is received, the devices save

---

[1]An exhaustive treatment of the communication between the FFMM classes and the actual devices can be found in [1].

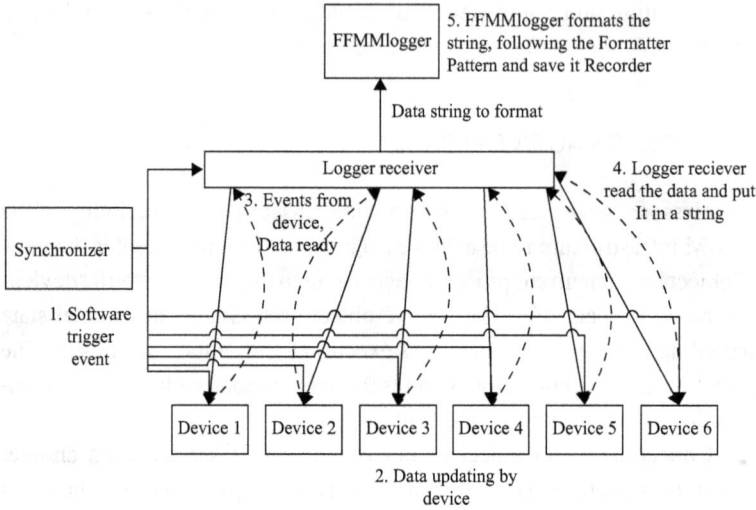

Figure 7.3. Active component design for logging infrastructure.

both their status and the measured data (e.g., a voltage value for the *Multimeter* or the position of the shaft from a *MotorController*). Finally, the *Logger Receiver* sends a message to interrogate the devices and retrieve the measurement data. Once collected, data are formatted in strings and saved by the *FFMMLogger* in a file or another stream.

7.2.3 CLASS TRANSDUCER

The class *Transducer* was designed as a specialization of the interface *Measurement Device*, common to all the devices used to handle a physical measurement instrument controlled directly from the PC through a communication bus (Figure 7.4).

Next, the use of the class *Transducer* in the application domain of superconducting magnets tests, namely the measurement of (a) the supply current and (b) the temperature of a Large Hadron Collider dipole at CERN, is highlighted.

For this purpose, specific methods are required in order to handle one or more multimeters.

7.2.3.1 Class Current Meter

The high intensity of the current flowing in the winding of a LHC superconducting magnet does not allow its direct measurement, and a

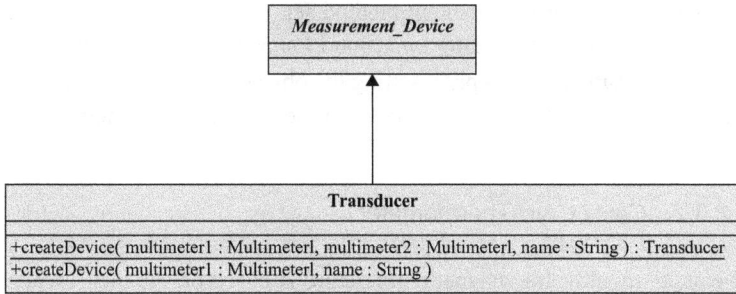

Figure 7.4. Diagram of the class *Transducer*.

high-accuracy Direct Current-Current Transformer (DCCT, [7]) has to be used. The primary current (equal to the current in the magnet) induces a current in the secondary DCCT winding, injected in a stable resistor to realize a voltage signal, where the signal is measured by a multimeter. The primary current is subsequently computed as the product of this voltage and the transduction factor, defined as the ratio between the maximal current on the primary and maximum value of the voltage output.

The class *Transducer* provides three methods to carry out the current measurement:

- *static Transducer* createDevice (std::string name,MultimeterI* mlVC)*
- *void setCurrentMeasurement(double Imax, double Vmax*
- *double getCurrent()*

The method *createDevice*[2] creates the *Transducer* object used to manage the measurement. Two parameters (*name* and *mlVC*) have to be provided. The *name* attribute uniquely identifies the object in the system. The second one is a generic object implementing the multimeter interface. The interface provides the transducer with the capability of using a multimeter object irrespective of the particular underlying physical device. An object, handling the multimeter connected to the DCCT, is passed to the *createDevice*. A reference to the transducer performing as the object *Current Meter* is returned.

---

[2]Each device in the FFMM project has a constructor method hidden in the *create-Device* in order to implement the singleton pattern: The instance of the device is identified by a unique name; if a *createDevice* is used with a name of an already present device, nothing is created and a reference to the existing object is returned.

The *setCurrentMeasurement* method sets the transduction factor for the considered DCCT, therefore *Imax* and *Vmax* have to be specified.

After creating and properly setting the object, the measurement of the magnet current is carried out by means of the *getCurrent* method.

7.2.3.2 *Class* Cryo-Thermometer

The assessment of the magnet temperature is based on the measurement of the resistance (R_{thm}) of a Resistance Temperature Detector (RTD), CER-NOX CX [8]. Figure 7.5 shows the schematic of the employed circuit. The sensor is placed inside the magnet on the steel collar, surrounding the superconducting coils. The upper block is configured to carry out a four-leads resistance measurement. The conditioning block generates three different excitation currents, *I*, through the sensor, depending on the resistance range. The conditioner provides two corresponding outputs: V_{ARI} and V_{out}. V_{ARI} is the Analog Range Indicator (ARI), indicating the resistance range, and V_{out} the output signal. The resistance of the sensor is finally computed as shown in Table 7.1.

Once the thermometer resistance R_{thm} has been calculated, the temperature can be obtained by substituting R_{thm} in the calibration curve:

$$T(R_{thm}) = \sum_{i}^{N} Z_i \ln\left(\frac{R_{thm}}{R_0}\right)^i \tag{7.1}$$

where the resistance R_{thm} of the sensor is expressed in Ω, R_0 is the resistance at 0°C, Z_i (i = 1, 2, ..., 9) are the calibration coefficients in K used to compute the curve [9], and *T* is the temperature in K. Usually, Equation 7.1 is provided in a tabular form, and the temperature is then computed via linear interpolation.

Two multimeters are used to measure the two voltages V_{ARI} and V_{out}.

Analogous to the *Current Meter*, the class *Transducer* defines three methods to implement the *Cryo-Thermometer* object:

- *static Transducer* createDevice(std::string name,MultimeterI* ml1Data,MultimeterI* ml2G*
- *void setTemperatureMeasurement(std::string fileName)*
- *double getTemperature()*

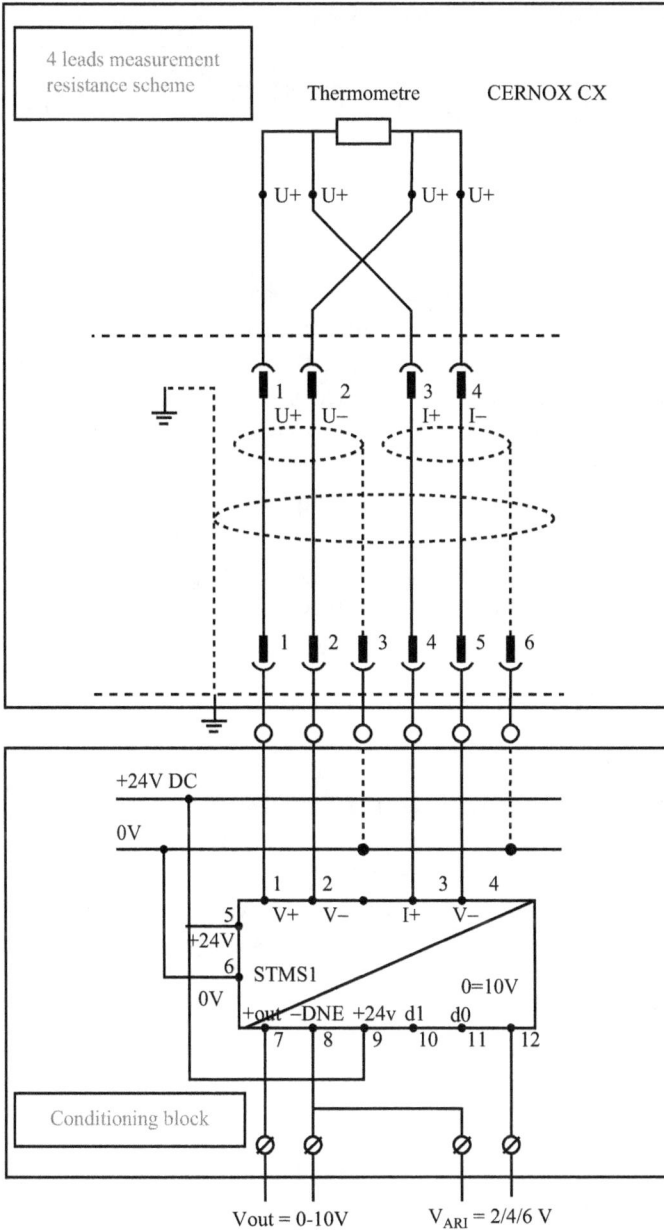

Figure 7.5. Circuit schematic of the physical cryo-thermometer RTD, CERNOX CX.

Table 7.1. Computation of the sensor resistance R_{thm}

	I	V_{ARI}	
$30\ \Omega < R_{thm} < 400\ \Omega$	100 μA	6 V	$R_{thm} = 40 \times V_{out}$
$400\ \Omega < R_{thm} < 4000\ \Omega$	10 μA	4 V	$R_{thm} = 400 \times V_{out}$
$4000\ \Omega < R_{thm} < 40000\ \Omega$	1 μA	2 V	$R_{thm} = 4000 \times V_{out}$

The *createDevice* method creates an object *Transducer* working as a Cryo-thermometer. It is an overloaded method. Its signature distinguishes it from the analogous method used to create a *Current Meter*. The parameter *name* identifies univocally the object in the system. Two generic multimeter objects, *ml1Data* and *ml2IG*, are needed in order to read V_{ARI} and V_{out}.

The *setTemperatureMeasurement* has as argument the name of the file containing a tabular representation of Equation 7.1, and loads the table.

The *getTemperature* carries out the temperature measurement by reading the V_{ARI} and V_{out}, computing R_{thm}, and interpolating the data stored in the table to obtain the magnet's temperature.

7.2.4 CLASS MIDIMOTORCONTROLLER

The class *MidiMotorController* is implemented in order to control a stepper motor SIMPA 1 AXE [9]. The controller provides the functions to drive the shaft of the rotating units used at CERN for the measurements based on rotating coils (Chapter 6, Section 6.2.1).

Figure 7.6 shows the diagram of the class *MotorController*. Its interface *IMotorController identifies* the methods common to different physical motor controllers.

The *MidiMotorController* implements the abstract class *IMotor Controller*, with a specific code capable of handling the particular underlying physical object. An object *MidiMotorController* is created by the following method:

```
static MidiMotorController* createDevice(std::string name, std::string
mod, std::string ser_num, std::string man, int reduction, int comPort).
```

The method requires the following parameters identifying the devices: *name* (unique name of the device), *mod* (model), *ser_num* (serial number), and *man* (manufacturer). The second to last parameter *reduction* depends

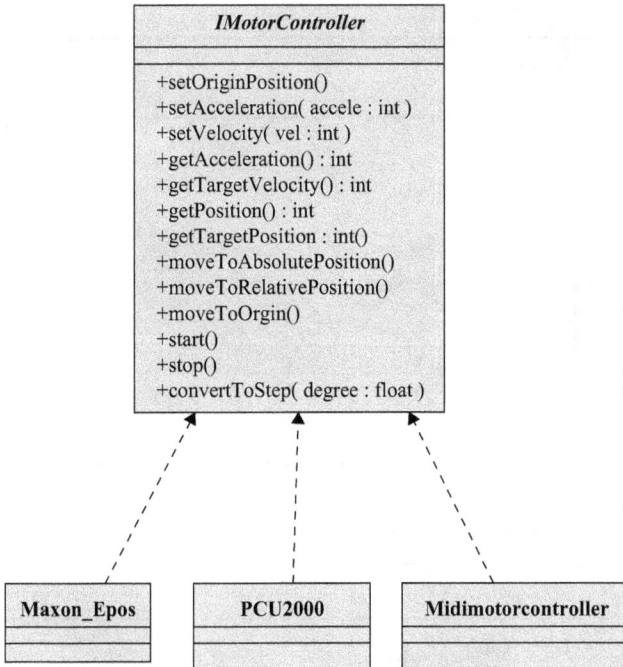

Figure 7.6. Diagram of the class *MotorController*.

on the particular gear mounted on the motor shaft, and, finally, *comPort* is the RS232 port address, connected to the remote controller.

The method *void EraseMotorMemory()* erases the controller memory in order to prevent instruction conflicts.

The interface *IMotorController* defines possible operation modes for the motors, while the method *void OperationMode(int OpMode)* sets the operation mode of the specified motor controller. Such a method is inherited and implemented by *MidiMotorController*.

In the following, the three operation modes are described. In velocity mode, the motor can turn in *continuous rotation*. The user can define a target velocity and the acceleration and deceleration profile used to reach it. An example is provided in Figure 7.7.

The following methods are available in velocity operation mode:

- *void setVelocity(int velocity)*: sets the velocity in rpm (revolution per minutes, positive in the clockwise direction or negative vice versa).
- *void setAcceleration(int Acceleration)*: sets the acceleration or the deceleration profile the motor has to follow to reach the desired velocity.

(a)

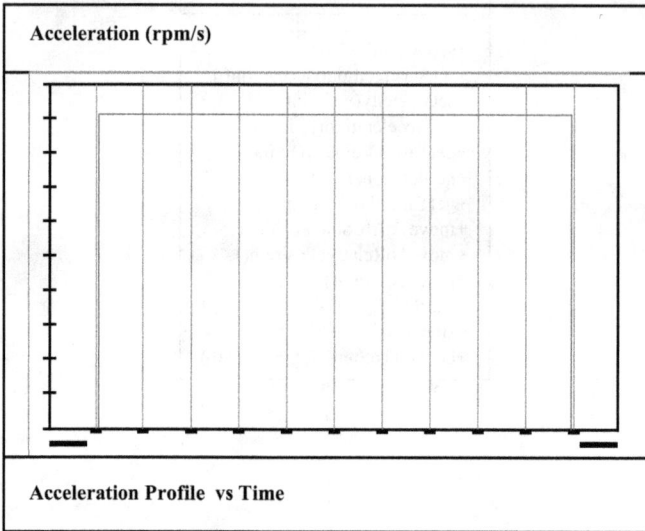

Acceleration (rpm/s)

Acceleration Profile vs Time

(b)

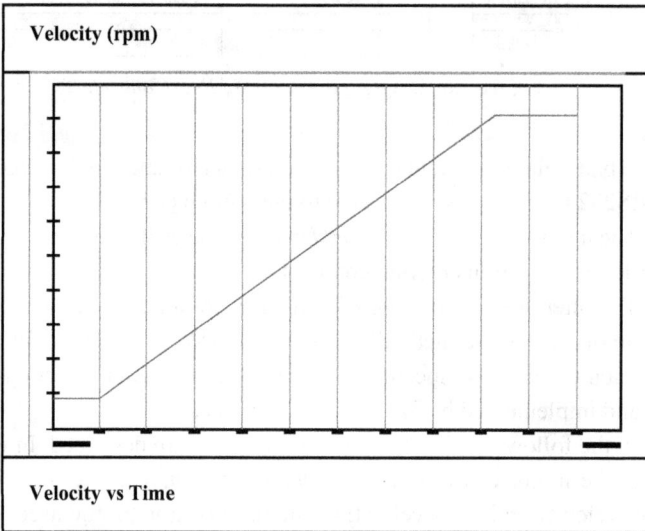

Velocity (rpm)

Velocity vs Time

Figure 7.7. Acceleration profile (a) and velocity profile (b).

- *int getTargetVelocity()*: returns the velocity set by *setVelocity* (rpm).
- *int getAcceleration()*: returns the acceleration set by *setAcceleration* (rpm/s).
- *void start()*: enables a continuous rotation with the desired target velocity and acceleration profile.
- *void stop()*: the motor is stopped with the desired deceleration.

The *set* and *get methods*, discussed for the velocity mode, are available also with the positioning mode operation. In this case, the motion of the motor starts and ends in a defined angular position, determined by an angular encoder. The positioning motion can be absolute or relative. In the former case, the end point is determined by an absolute position value defined with respect to a zero position, identified by means of a home sensor. In the relative position mode, the end point is defined with respect to the current position.

The following methods are used for the absolute and relative positioning:

- *void moveToAbsolutePosition(int step)* moves the axis to the absolute position referred to a null point, whose default value can be changed through the *OriginPosition()*[3] method. The parameter value is the number of steps. As an example, for the motor considered, a whole turn is completed in 1611 steps.
- *void moveToRelativePosition(int step)* moves the axis of the given number of steps starting from the current position. The *step* value can be either positive or negative according to the direction. To simplify the motor shaft positioning, a method to convert angular position in degrees to the corresponding number of steps was implemented.
- *int convertToStep(float degree)*: returns the value of step corresponding to the degree value.

7.2.5   *CLASS* Fast Digital Integrator

The class *FastDI* allows the remote control of the Fast Digital Integrator (FDI) by means of the PXI communication bus [10]. The files *FastDI.h* and *FastDI.cpp* contain the definition of the class and the implementation of their methods, respectively. In the file headers, all the methods are described, by specifying the functions, the input and output parameters, and the method preconditions. In the following, the main functions of the class are outlined.

The constructors are hidden in the methods *createDevice*. In particular:

- *static FastDI* createDevice(std::string name)* is the default function; and

---

[3]*setOriginPosition()*, defined in the interface and implemented in the *MidiMotorController* class, sets the actual position as zero reference for the motor controller.

- *static FastDI* createDevice(std::string name, std::string mod, std::string ser_num, std::string man)*, where
  - o *name* is the name of the device;
  - o *mod* is the device model;
  - o *ser_num* is the device serial number; and
  - o *man* is the device manufacturer.

  creates the device FDI and specifies its communication parameters.

The following methods specify the communication bus parameters and open the communications:

- *void configure(int bus, int slot, U8 barindex = 2, bool remap = 0, ACCESS_TYPE ac = BitSize32, int to_meas = 25)*
- *int Set_Communication_Bus(int bus, int slot, U8 barindex = 2, bool remap = FALSE, ACCESS_TYPE ac = BitSize32, int to_meas = 25)*

By means of the following parameters:

- *bus*, specifying the bus position in the PXI crate
- *slot*, specifying the slot position in the PXI crate
- *barindex*, a communication parameter with default value 2
- *remap*, controlling the virtual remapping; remap = 0 implies no virtual addressing (default value)
- *ac*, the access type on the bus
- *to_meas*, specifying the maximum time interval between two measurement points. It must be set according to the trigger frequency
- int Close_Communication_Bus(), closing the communication bus
  - o static void deleteDevice(std::string name), deleting the device

The parameters of the devices are set for the desired type of measurement by means of the following methods:

1. *int setGain(double gain)*, setting the gain value
2. *int set_ADC_Rate(int rate)*, setting the ADC sampling
3. *int setSamplesNumber(int sample)* sets the parameter "sample" (sample number) and specifies the number of samples to be acquired in case of a finite acquisition by setting the internal parameters *NumberofTurn* and *Number_of_trigger_per_turn*. The product of those parameters is the sample number to be acquired. *NumberofTurn* is the register of FDI containing number of turn for a finite acquisition (as a power of two). *Number_of_trigger_per_turn* is the number of trigger in each turn

4. *int Set_Number_of_Turns(int sample)*, setting the number of samples per turn

5. *void setTimeoutMeas(int timeout)* setting the timeout measurement, defined as the maximum time to waiting for the start of measurement

6. *int Set_Number_of_Trigger_per_Turn(int trigger)*, setting the number of triggers per turn and establishing the total number of samples to be read in case of finite acquisition

7. *int set_Buffer_Acquisition_Size(int size)*, setting the FDI's parameter *BufferSizeAcquisition* and specifying the block length of the FDI memory to be read. The size value must be a power of 2 and cannot be greater than the half of the FDI memory size (4096 word of 32 bit)

8. *int set_Reading_Number_Per_Event(int events)*, setting the parameter Reading_Number_per_Event and specifying the number of values to be read for each trigger pulse

The set methods are summarized in the method *setFDI(int sampleNumber, double gain, SAMPLE_MODE mode, double rate, int buffsize, int timeoutMeas, int readEvent)*, which applies all the main settings of the instrument.

The following methods sets the FDI:

- *int getADCRate(double *rate)*, saving in "rate" the ADC sampling rate
- *int getGain(double *gain)*, saving in "gain" the FDI gain
- *int getSampleNumber(int *sampleNumber)*, holding in "samplenumber" the parameter set by "setSampleNumber"
- *int getTimeoutMeas () const*, holding in "timeout" the value set by "setTimeoutMeas"
- *int Read_Number_of_Turns (int*)*, getting the number of samples per turns
- *int Read_Number_of_Trigger_per_Turn(int*)*, setting the number of trigger events per turn
- *int Read_Buffer_Acquisition_Size (int* buffsize)*, saving in "buffsize" the value set by "int set_Buffer_Acquisition_Size"
- *int Read_Reading_Number_per_Event (int* event)*, holding in "event" the value of the FDI parameters Reading_Number_per_ Event

The FDI is calibrated by means of the following functions:

- *int Calibrate(double gain)*. It calibrates the gain specified as input parameter. If the input parameter is −1, then the calibration of all the gains is launched. The method called without arguments launches the calibration for the current gain.

- *void Calibrate_Cmd ()*, sending the calibration command.
- *int Calibrate (void)*, launching the calibration for the current gain.

Capabilities of status check are provided in order to control the operation of the device. The method used to read the status register is:

- *int Read_state(int *state)*, which reads the instrument status register and saves the value in the argument pointer.

The acquisition procedure consists of four steps by the following functions:

- *void start()*, which puts the FDI in Measurement acquisition waiting for the first trigger.
- *int PollingDataready()*: It waits for the set of the bit dataready, it returns 0 if the operation was successful.
- *int ReadBlock (int Address, float * pData)*, reading a memory block of the size specified by bufferAcquisitionsize (32 bit) after the polling on the dataready bit. Address is the starting address, pData is the pointer for the data.
- *int ReadBlock2 (int Address, void *pData)*, reading a block memory of the size specified by bufferAcquisitionsize (32 bit). Polling is not done in this function. Address is the starting address, pData is the pointer for the data.
- *Int stop()*, stops the acquisition if a continue acquisition is set. If the acquisition is a finite acquisition, the stop has no effect, because the measure procedure stops when the given *sample number* is reached.
- *int Finite_Acquisition (std::string file_name)*, handles an acquisition of a finite number of samples previously specified. It manages also the start of the acquisition and data saving. It is a blocking function, so it will be called in a separate thread. It's arranged in three parts: start, polling, and read Block. Until the number of sample (set by *setNumberOfSample*) is reached it calls the *pollingReady()*. If the return value of the *pollingReady()* is zero, it invokes *readBlock()*.
- *int Stop_Acquisition ()*, stops the current acquisition.

When an acquisition starts and stops, the FDI sends an event to the listeners. These events (start and stop event) don't contain any data. The read block event informs the framework about the reading of a FDI's memory block. The generated event object contains the read data, a pointer to the current read block, and the number of remaining blocks. The following method is used to generate timeout event:

- *int FDITimeout(time_t tstart, int timeout)* detects a timeout condition; it starts counting from *tstart*; if *m_timeout* is exceeded the function returns 1, otherwise 0. If a timeout elapses, an error event will be launched containing the timeout reached, a time out condition can be detected by *FDITimeout*, when the block is due to an internal error of FDI, but the same error will be launched consequently to a communication bus timeout.

7.3   CORE SERVICE LAYER

7.3.1   FAULT DETECTOR

In this section, the classes implementing the service of fault detection are detailed. The objective of such classes is to handle device faults. This service is hidden to the FFMM user. As in the Event Handling service, the fault detection strategy uses the *Event_info* objects to store information about Fault Events. For each device, the *Event_info*, *fault_info*, is defined. In this case, the *Event_info* is composed of four nodes:

- "DEV_TYPE": identifies the device
- "SENDER_METHOD": the method of device that threw the fault
- "STATUS": the device status
- "ERROR": the error type

Each Fault Event notification can be either preceded or followed by *fault_info* as in the following:

fault_info.erase_insert("SENDER_METHOD","NameMethod")
fault_info.erase_insert("STATUS", "NameStatus")
fault_info.erase_insert("ERROR","Message Error")
name_fault.notify(this,fault_info)

The structured information is used by the *Fault Detector* to know exactly where the Fault Event was notified, and how to react.

Then, in the following section, the implementation of the *Fault Detector* class and the *IFault* interface are shown.

7.3.1.1 *Class* FaultDetector

The fault detection policy is based on the communication of Fault Events from the device involved in a measurement application to the *Fault Detector*. The reaction to a faulty condition consists of (a) adding the *Fault Detector* as listener of the Fault Events, and (b) for each fault category, implementing a reaction method (invoked at the occurrence of a fault).

The constructor method is used in a macro DEVICE_CREATION inside the method script defined in the file ffmm.h:

- *Fault_Detector(TestManager* tm)*
  - o In such a way, the Fault Detection service in FFMM is hidden to the final user. About reaction methods, different types of Faults are available: Fatal, Configuration, Warning, and Local, with corresponding distinct reaction methods. Methods to add and remove listener to Fault events are defined too. The architecture of the *Fault Detector* is analogous to the *Event Handling*. In fact, the method adding and removing are implemented by using the *Delegate* by POCO libraries.

In the following, the implementation of the fault methods is illustrated.

Fatal Fault Methods

The occurrence fault classified as fatal involves the end of the measurement operations. For this reason, the *Fault Detector* policy is centralized in the following methods:

- *void addOnFatalFault(vector<Virtual_Device*> vl).*
  /*Precondition: An array of all the registered devices has to be available (referring to the class *Virtual Device*). Postcondition: On all the registered devices, the *Fault Detector* is added as listener of the event fatal fault.*/ The method *addOnFatalFault* enables to handle the fatal faults that can occur in a measurement session. The argument is the array of the registered *Virtual_Device*, which contains the pointers to the overall Devices involved in the measurement application.
  When a *Fatal Fault* occurs, the *Fault Detector* calls the reaction method:
- *void reactFatalFault(const void* pSender, Event_info& arg)*;

The arguments "pSender" and "arg" are the sender device and the fault info, respectively. The method *reactFatalFault* ends the application, after which all the devices are brought back to the safe state. Each device implements the response to a *Fatal Fault* via the method *onFatalFault*, called by the *Fault Detector*.

Configuration Fault Methods

A *Configuration Fault* can occur during the devices configuration phase.

- *void addOnConfigurationFault()*
  /*Precondition: A vector of all registered devices has to be available (that vector has to be set by addOnFatalFault before using the method). Postcondition: on all the registered devices the Fault Detector is added as unique listener of the Event Configuration Fault*/ Such a method is used in the macro DEVICE_CONFIGU-RATION inside the script method defined in the ffmm.h header file. About the reaction method to Configuration Fault, the FAULT DETECTOR calls the methods *onConfigurationFault* implemented in each devices.
- *void reactConfigurationFault(const void* pSender, Event_info& arg)*
  The method aims at reacting to a wrong state of the device bus open connection operation.

Warning Fault Methods

The warning fault can occur during each phase of a measurement application. The following method enables to handle the Warning fault:

- *void addOnWarningFault( )*
  /*Precondition: a vector of all the registered devices has to be available (the same vector set in *addOnFatalFault()*. Postcondition: for all the devices, the Fault Detector is added as listener of Warning Fault.*/ In the ffmm.h header file, this method is used in a macro SET_DEVICE. Different from other faults, the reaction method calls other private methods, which are implemented in the class *Fault Detector*. The response depends on the device.
- *void reactWarningFault(const void* pSender, Event_info& arg).*
  The Fault Detector manages the Warning Fault Occurrence using other specified methods specialized for each device and called in

the reaction methods. In particular, only the reaction to warning fault was directly implemented: more sophisticated behaviors were not needed.[4] Private methods are implemented in order to provide better warning handling:

o *void FDI_warningFault(Virtual_Device* dv, Event_info& arg)*
o *void FDI_Cluster_warningFault(Virtual_Device* dv, Event_info& arg)*
o *void EncoderBoard_warningFault(Virtual_Device*dv, Event_info& arg)*
o *void MidiMotorController_warningFault(Virtual_Device* dv, Event_info& arg)*
o *void Maxon_Epos_warningFault(Virtual_Device* dv, Event_info& arg)*
o *void Keithley2k_warningFault(Virtual_Device* dv, Event_info& arg)*
o *void LVPowerSupply_warningFault(Virtual_Device* dv, Event_info & arg)*

In such a way, each device has a different warning fault policy. The policy concerns in particular the logging of messages.

Local Fault Methods

The Local Fault notifies a situation of a component in a not recoverable faulty status. In this case, the component is excluded from the application. The method that defines the listener of this fault is:

• *void addOnLocalFault().*
  /* Precondition: a vector of all the registered devices has to be available (the same vector set in *addOnFatalFault*). Postcondition: on all the devices the Fault Detector is added as listener of Local Fault*/. In the ffmm.h header file, this method is used in the macro SET_DEVICE. In order to properly react to a local fault, the *Event_info* has to be filled with the following format: "Device_Type," that identifies the device; "Method," the method of the device that threw the fault; and "fault ID," fault identification code. Then the reaction method is defined as:
  o *void reactLocalFault(const void* pSender, Event_info& arg).*

---

[4]The other fault types are handled within the reaction method.

7.3.1.2 Interface IFault

The interface *IFault* provides the prototypes of the methods used by the Fault Detection service. This interface is inherited by all the devices. The concrete implementation of the interface is carried out inside the devices. In the following, the *IFault* interface is shown:

- *virtual void onFatalFault( ){}*
- *virtual void onConfigurationFault( ){}*
- *virtual void onWarningFault( ){}*
- *virtual void onWarningFault(int level){}*
- *virtual void onLocalFault( ){}*

The method *onFatalFault* is the primary response to Fatal Fault, and it is implemented in order to stop the devices and the application. Inside of every device, the *onConfigurationFault* is implemented in a similar way; it deals with the configuration of a device. In regards to the *onWarning-Fault*, there are two versions of prototypes. In case of a Warning Fault, the response is usually the logging of messages to console, to file, or both. Dealing with the *onLocalFault*, not all the devices implement the method. The method either excludes the device from the application or carries it toward the ready state. The remaining method is:

- *virtual Event_info onCleanFault(){Event_info inf; return inf;}*

In all the devices, this method is used for cleaning a device ad state (from a previous anomalous application end).

7.4 MEASUREMENT SERVICE LAYER

7.4.1 SYNCHRONIZER

In the following, the implementation of the Petri net's (PN) based components, the classes *Labeled Petri Net* (LPN) and *Synchronizer,* are described. The basic idea is to have a software component being able to manage the execution of several generic tasks.

The *Synchronizer* illustrated here is based on the concept of an Execution Graph (Figure 4.12, Chapter 4). The Execution Graph is a set of nodes (task or event), and arrows interconnecting them. To model an Execution Graph, the *Synchronizer* uses the object PN. There are two types of node: *Place* and *Transition*. Each node is connected via an oriented arc (Figure 4.12, Chapter 4). Therefore, *Place, Transition*, and *Arc* are the basic components of a PN. As described in Chapter 4, an LPN is proposed for extending the PN. In the LPN, each *Place* and *Transition* has an associated label in order to allow the definition of different classes for both of them.

7.4.1.1   Basic Petri Net Component

In the following, the implementation of the classes and the interfaces for (a) *Place*, (b) *Transition*, and (c) *Arc* are described.

* **Place**

   The *Place* object represents the condition status of a node. Each Task Node has three *Places*, defining the three available states: EXECUTE, FREEZE, and TERMINATE, while each Event Node has one *Place*, TRIGGERED. The *Places* can contain *tokens*; the current state of the modeled system is given by the number (and type if they are distinguishable) of tokens in each place. This operation is called "marking" of tokens. The constructor method is:

   o *Place(string id, string label, int tokens)*

   The *string id* defines a member of a node, the *string label* identifies the *Place* type within the node and the *int tokens* is the value giving the number of tokens inside *Place*. In this class, two private vectors are defined:

   o *vector<Arc*>   inArcs*
   o *vector<Arc*>   outArcs*

   useful to store all the *Arcs* objects that entry and exit from a *Place*. For this, the methods Add(Arc) and Get(Arc) are implemented:

   * *void addInArc(Arc* inArc)*
     o *void addOutArc(Arc* outArc)*
     o *vector<Arc*>   getInArcs(void)*
     o *vector<Arc*>   getOutArcs(void)*

* **Transition**

   The object *Transition* represents the events or actions that can occur in a graph between two *Places*. *Transitions* are active components, defining the two Events of Task Node, START and STOP, and one event of the Event Node, TRIG. They model the activities (the transition *fires*), thus changing the state of the system (the marking of the PN). *Transitions* are only allowed to fire if they are *enabled*, which means that all the preconditions for the activity must be fulfilled (there are enough tokens available in the input *Places*). When the transition fires, it removes tokens from its input *Places* and adds some of them at all of its output *Places*. The number of tokens removed and /added depends on the cardinality of each *Arc*. In the constructor method:

- *Transition(string id, string label, int priority, bool enable = true)* the *string id* defines the node of membership, the *string label* identifies the *Transition* type within the node, the *int priority* establishes if one *Transition* has priority, and the *bool enable* defines if the *Transition* is enabled. The following methods are implemented:
  - o *bool isEnable(void)*, verifying if the transition is enabled
  - o *void setEnable(bool enable)*, setting the enabled transition
  - o *bool tryToEnable(void)*, it trying to enable the transition

  Such as the class *Place*, two vectors to store the *Inner Arc* and *Outer Arc of Transition*, and the methods *Add(Arc)* and *Get(Arc)* are implemented.

- **Arc**

  The object *Arc* represents the connection between *Places* and *Transitions*. *Input Arcs* connect *Places* with *Transitions*, while *Output Arcs* start at a *Transition* and end at a *Place*. In the constructor method:
  - o *Arc(string id, string label, Place* place, Transition* transition, int weigh = 1)*,

  the *string id* defines the node of membership, the label identifies the *Arc* type, Inner or Outer, place and transition are pointers at *Place* and *Transitions*, respectively; and the integer weigh is the cardinality of the *Arc* (this establishes the number of tokens removed and added). Also in this case, Set and Get methods are implemented for these arguments.

7.4.1.2 *Class* Labeled Petri Net

PN is a formalism particularly suited for asynchronous and parallel system specification and analysis. The basic feature is the capability of simultaneously presenting control and data flows in concurrent systems. The graphical representation of the network is a dual graph containing two types of nodes, *Places* and *Transitions*. The nodes are connected by directed Arcs. An Arc cannot connect the nodes of the same type. *Places* in the network contain tokens and are represented as circles or ellipses (states), while

Transitions are displayed as rectangles (simulating events). Classical PNs are extended as LPNs in order to offer a more consistent description, and to simplify modeling and analysis. In an LPN, each *Place* or *Transition* has an associated label to model, in addition to, managing different classes. In the following, the implementation of LPN class is shown for (1) *Place*, (2) *Transition*, and (3) *Arc* methods.

Place Methods

The following Place Methods are implemented to insert and carry out a *Place* in the reference graph:

- *Place* addPlace(string id, string label, int nTokens)*: It adds a new place with specified id (identification nod), label (identification type of Place), and number of Tokens. The label string can assume three values in Task Node (EXECUTE, FREEZE, and TERMINATE), and one value in Event Node (TRIGGERED).
- *Place* getPlace(string id, string label)*: it looks for a Place with specified id and label (to see addPlace).

Transition Methods

Analogously, add and get Methods are implemented:

- *Transition* addTransition(string id, string label, int nTokens)*: It adds a new *Transition* in the graph. The variable id is the identification Node, label is the type of *Transition*, and nTokens is the number of Tokens. The label string can assume two values in a Task Node (START and STOP), and one value in an Event Node (TRIG).
- *Transition* getTransition(string expectedTransId, string expectedTransitionLabel)*: it looks for a Transition with specified id and label (to see *addTransition*).

The following methods are used to get information about the state of a *Transition*, that is, if it is enabled or disabled. Both the methods use *getTransition* to identify the specified *Transition*, and they check the state by means of method *isEnable* (*Transition* class).

- *Transition* getTransitionEnabled(string expectedTransId, string expectedTransitionLabel)*
- *Transition* getTransitionDisabled(string expectedTransId, string expectedTransitionLabel)*

Moreover, the last method about *Transition* implements the possibility to check all the *Transitions* enabled in a PN:

- *vector<Transition*> getTransitionsEnabled(void)*

Regarding the *Transition* of type TRIG, the following method was implemented, which is usually used when a *Transition* is to be triggered with another one (Event Node):

- *bool trigTransition(string transitionId, string transitionLabel, string transitionOutArcId = " ", string transitionOutArcLabel = " ")*

The implementation of the precedent method follows these steps:

- Verify that transition is enabled and disable it
- Remove *n* tokens from previous places (with *n* = in arc *weigh*)
  - o Select one arc to pass through
  - o Select the first out arc (lazy trigger)
- Select the required out arc (deterministic trigger)
- Add *n* tokens to the place pointed by selected out arc (with *n* = out arc *weigh*)
- Enable transition with in-arc weigh greater of in-place tokens

Arc Methods

Arc Methods manage all the connections between *Places* and *Transitions*:

- *Arc* addPlaceToTransitionArc(string id, string label, Place* place, Transition* transition, int weigh = 1)*: it defines a new Arc from *Place* to *Transition*;
- *Arc* addTransitionToPlaceArc(string id, string label, Transition* transition, Place* place, int weigh = 1)*: it defines a new Arc from *Transition* to *Place*.

7.4.1.3 Class Synchronizer

The class *Synchronizer* is aimed at building the Execution Graph, by adding a node, an event, or an arrow, with the possibility of querying the Execution Graph, by determining the executable nodes, the end node, and the loop detection. Moreover, another feature is the capability of updating the graph, by forcing the execution dynamics: execute, terminate, or freeze a node, notify an event, and so on. To achieve this objective, the class *Synchronizer* uses LPN, *Transition*, *Place*, and *Arc* classes.

In the ensuing text, the implementation of the *Synchronizer* is presented. The constructor method is:

- *Synchronizer(void)*: it creates a new *Petri Net* object, and the destructor method
- *~Synchronizer(void)*: it deletes the *Petri Net* object.

The methods are divided into three groups: Task Node, Event Node, and Node Status.

Task Node Methods

In succession, the methods for adding a Task Node in the Execution Graph are:

- *void addRootNode(string nodeId)*: it adds a Root Task Node to the Execution Graph, with nodeId defined.
- *void addInnerNode(string nodeId)*: it adds an Inner Task Node;

A Task Node has three possible *Places* (states) and two possible *Transitions* (events). In the implementation, the first two methods, *addRootNode* and *addInnerNode* use a private method, *addNode*. The following method has the aim of defining an Arrow (connection) between two Task Nodes:

- *void addTaskToTaskArrow(string idSrcNode, string idDstNode)*

It takes like arguments, the *idSrcNode* string of Source Node, and the *idDstNode* string of Destination Node.

Two Get methods are implemented to know which Nodes are executable, and the end nodes to execute:

- *vector<string> getExecutableNodes(void)*: It gets a vector with all the executable nodes. Nodes are executable when their transitions are with "START" label.
- *set<string>* getEndNodes(void)*: it gets all end Task Node.

Event Node Methods

In the following, the entire Event Node methods are described. An Event Node is made up of two elements, a *Place* (TRIGGERED) and a *Transition* (TRIG). The following method:

- *void addEventNode(string genericEventId)*
adds an Event Node. The following method defines an Arrow (connection) between a Task Node and an Event Node:
- *void addEventToTaskArrow(string idSrcGenericEvent, string idDstNode).*

It takes like arguments, the *idSrcGenericEvent* string of Source Event Node, and the *idDstNode* string of Destination Task Node. For the Event Node, another method:

- *bool notifyEvent(string nodeId)*

is used to notify an Event. The Event notification gives the start to TRIG Transition of Event Node. This mechanism allows the execution of the Task Nodes connected to Event Node.

Node Status Methods

The following methods are implemented for managing and checking the Node Status, by allowing a Node to be set in a particular state. The status methods are as follows:

- *void setNodeExecutable(string nodeId)*: It sets a Task Node in Executable status. This method enables the START Transition in the node to make the EXECUTE Place available.
- *bool execute(string nodeId)*: It sets the nodeId Node in EXECUTE Place, throwing the trigTransition.
- *bool terminate(string nodeId)*: It sets the Node in terminate status. This method enables the STOP Transition that makes available the TERMINATE Place, by means of STOP_TERMINATE Arc.
- *bool freeze(string nodeId)*: It sets the Node in freeze status. This method enables the STOP Transition that makes available the FREEZE Place by means of STOP_FREEZED Arc.
- *bool unFreeze(string nodeId)*: It sets the Node in unfreeze status. This method enables the START Transition that makes available the EXECUTE Place.

The last method is used to check the status of the *Execution Graph*, by checking the possible Loop status, that is, the lack of End Nodes:

- *bool checkLoop(void)*

7.5 USER SERVICE LAYER

7.5.1 MEASUREMENT DOMAIN SPECIFIC LANGUAGE

The first part of this section shows how to create a new project in Eclipse by using the plugin *openArchitectureWare (oAW)* in order to define a new DSL. The second part shows how the new language Measurement Domain Specific Language (MDSL) is implemented.

7.5.1.1 Eclipse Platform

The free and open source software platform Eclipse is exploited for DSL development. Eclipse is a multilanguage platform that includes an integrated development environment (IDE) and a plugin system for its extension. It is written primarily in Java and is used to develop applications using the same language, but by means of various plugins, other languages can also be utilized in code writing, namely, C, C++, COBOL, Python, Perl, PHP, and so on.

Eclipse employs plugins in order to provide all of its functions on top of (and including) the runtime system, in spite of some other applications where the function is typically hard coded. The runtime system of Eclipse is based on Equinox, an OSGi standard compliant implementation. This plugin mechanism is a lightweight software componentry framework.

The key to the seamless integration (but *not* of seamless interoperability) of tools with Eclipse is the plugin. Except for a small run-time kernel, everything in Eclipse is a plugin. This means that each developed plugin integrates with Eclipse exactly as other plugins; in this respect, all features are created equal. Eclipse provides plugins for a wide variety of features, some of which arise from third parties using both free and commercial models. Examples of plugins include UML plugin for Sequence and other UML diagrams, plugin for Database explorer, and many others. The Eclipse SDK includes the Eclipse Java Development Tools, offering an IDE with a built-in incremental Java compiler and a full model of the Java source files. This allows for advanced refactoring techniques and code analysis. The IDE also makes use of a *workspace*, in this case a set of metadata over a flat filespace allowing external file modifications as long as the corresponding workspace "resource" is refreshed afterwards. Eclipse's widgets are implemented by a toolkit for Java called Standard Widget Toolkit (SWT), unlike most Java applications, which use the standard Abstract Window Toolkit (AWT) or Swing. Eclipse's user interface also uses an intermediate GUI layer called JFace, which simplifies the construction of applications based on SWT.

openArchitectureWare

When starting a new project in order to define a new language, first a *xText* project must be created. In this book, *xText* projects are based on the Eclipse plug-in architecture openArchitectureWare (oAW). This section aims at illustrating the definition of external DSLs using tools from the Eclipse Modeling Project (*EMP*).

oAW is currently one of the most used frameworks. Much of this success derives from its flexibility: Rather than providing premade generator templates, oAW serves as a generator toolkit and enables users to easily create tailored generator solutions that really fit their needs. Besides this flexibility, oAW users benefit also from the tight integration with Eclipse: Not only does oAW come with an array of editors that make writing templates and workflows an easy task but also delivers refactoring support, easy navigation, an incremental project builder, and a debugger. It supports parsing of arbitrary models, and a language family to check and transform models, as well as generate code based on them. Supporting editors are based on the Eclipse platform. oAW has strong support for Eclipse Modeling Framework (EMF)-based models but can work also with other models, for example, UML2, XML, or simple JavaBeans. At the core, a workflow engine allows the definition of generator and transformation workflows. A number of prebuilt workflow components can be used for reading and instantiating models, checking them for constraint violations, transforming them into other models, and, then finally, for generating code. In other words, oAW helps with meta-modeling, constraint checking, code generation, and model transformation.

More recently, the framework *xText* has been developed to support the creation of textual DSLs. First, the DSL is defined in an *xText* grammar, then the *xText* framework is exploited to generate a parser, an *Ecore-based* meta-model, and a textual editor for Eclipse. Afterwards, the DSL and its editor are refined by means of *xTend* extensions. Finally, the template language *xPand* is exploited to generate code out of textual models. The actual content of this example is rather trivial: The DSL will describe entities with properties and references between them from which Java classes are generated according to the JavaBean conventions, a rather typical data model. In an actual scenario, also persistence mappings and so on can be generated from the same models.

xText Project

xText is part of the oAW project, which is in turn part of Eclipse Generative Modeling Technologies (GMT). Based on an Extended Backus–Naur Form (EBNF) like notation, *xText* generates the following artefacts:

- A set of Abstract Syntax Tree (AST) classes represented as an EMF-based meta-model.
- A parser that can read the textual syntax and returns an EMF-based AST (model).
- A number of helper artifacts to embed the parser in an oAW workflow.
- An Eclipse editor that provides syntax highlighting, code completion, code folding, a configurable outline view, and static error checking for the given syntax.

xText starts from a description of a textual syntax (the grammar) and derives an AST class model (the meta-model) from that concrete syntax definition. Cross references within the same model or through different models can be linked separately from the textual syntax description. Linking can be a quite complicated process if scopes, namespaces, and visibility of elements are considered: It is crucial for a textual language framework to allow parsing and linking to be separated. Parsing and linking separation helps to implement more sophisticated linking logic independently of the concrete syntax. Additionally, the AST can be checked before doing additional linking and transformations. In some cases, the user doesn't even want to link references up-front, but wants them to be looked up dynamically.

Linking in *xText* can be done in several ways. The easiest way is to make use of so called extensions, that is, operations that can be annotated to existing meta-classes. Another solution is to transform the AST to an "actual" meta-model. This has the additional advantage that the concrete syntax can be changed, or several different concrete syntaxes can be created for the same meta-model. The necessary transformation is relatively straightforward to define, because it is basically a one-to-one mapping with some additional linking logic.

To create a new textual DSL with *xText*, up to three files that depend on each other are needed, according to the following steps:

- Start up Eclipse with oAW installed in a fresh workspace.
- Select **File > New... > Project... > openArchitectureWare > Xtext Project.**

- Specify the project settings in the wizard dialog.
- Click **Finish** (Figure 7.8).

The wizard creates three files, *my.dsl*, *my.dsl.editor*, *and my.dsl. generator*:

- **my.dsl** is the language project, where the grammar for the DSL is created. After running the Xtext generator, this model also contains a parser for the DSL and a meta-model representing the language.
- **my.dsl.editor** will contain the DSL editor.
- **my.dsl.generator** contains an oAW code generator skeleton.

Defining the Grammar

An *xText* grammar consists of a number of rules (Model, Message, Field, and Type). A rule is described using sequences of tokens. A token is either a reference to another rule or one of the built-in tokens (STRING, ID, LINE, and INT). *xText* automatically derives the meta model from the grammar. Conversely, the meta model is basically a data structure whose instances represent the structure of sentences in the language.

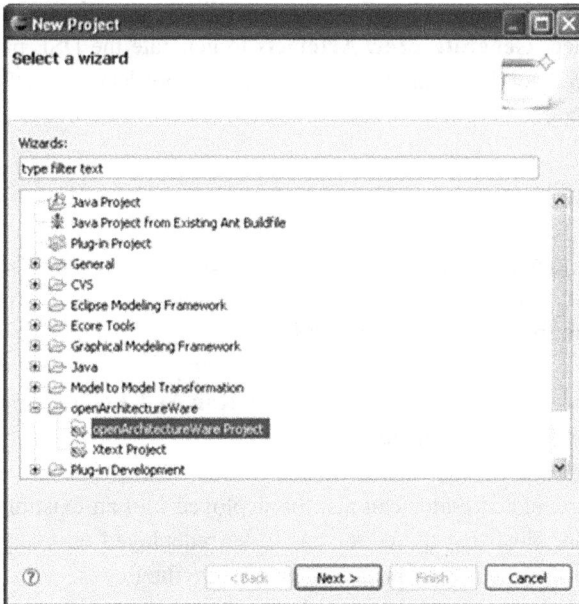

Figure 7.8. Wizard to start new *Xtext* project.

A rule results in a meta type, and the tokens used in the rule are mapped to properties of that type (comments, name, and fields). Different assignment operators are used. The equals sign (=) just assigns the value returned from the token to the respective property (the property will have the type of the token) and "+=" adds the value to the property.

Thus, after creating the new *xText* project, the grammar is created (an example is shown in Figure 7.9). The grammar specifies the meta-model and the concrete syntax for the desired MDSL.

7.5.1.2 The MDSL Editor

For the MDSL editor, the grammar language provided by *xText* is exploited. The following screen shots in this section show how the syntax is described for the FFMM-DSL. In fact, language and tooling used for describing the DSL syntax are bootstrapped, that is, they are implemented using the *xText* framework itself. Bootstrapping is a common technique in the field of language and compiler development. If language and tools can be bootstrapped, this proves a certain level of maturity of the tools.

After specifying the grammar, the DSL editor can now be generated:

- Right-click inside the *xText* grammar editor to open the context menu.
- Select **Generate *xText* Artefacts** to generate the DSL parser, the corresponding meta-model and, last but not least, the DSL editor (Figure 7.10).

Running the Editor

To see the generated editor in action, the plugins must be run in an Eclipse installation. The most convenient way to do this is to start a new Eclipse application from within the running Eclipse:

- Select the editor plugin and choose **Run As > Eclipse Application** from its context menu.

The generated editor can also be deployed into an existing Eclipse installation. Note that the editor has to be redeployed on every change applied to the plugins. To install the editor into the Eclipse currently running, the following steps are needed:

Figure 7.9. DSL grammar.

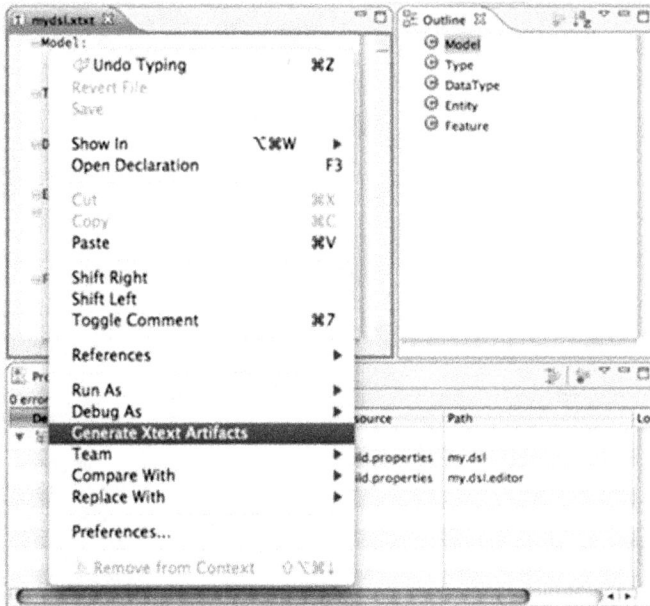

Figure 7.10. Generate *Xtext* artefacts.

- Choose **Export... > Deployable plug-ins and fragments...**
- The *Export* dialog appears. Select the three DSL plugins.
- Enter the path of the Eclipse installation. Make sure the selected directory contains the Eclipse executable and a folder named *plugins*. Usually, the directory is called *eclipse*.
- Choose **Finish** (Figure 7.11).
- Restart Eclipse.

Code generation With *xPand*

The *xText* wizard already created a generator project. In this design, the FFMMs class is shown to be connected with the new language DSL.

Part of the implemented *xPand* is shown in Figure 7.12.

The Grammar Language

At the heart of *xText* lies its grammar language, like an extended Backus–Naur Form (BNF), but it doesn't describe only the concrete syntax, but can be also used to describe the abstract syntax (meta-model).

As stated before, the grammar is not only used as input for the parser generator, but it is also used to compute a meta-model for the DSL.

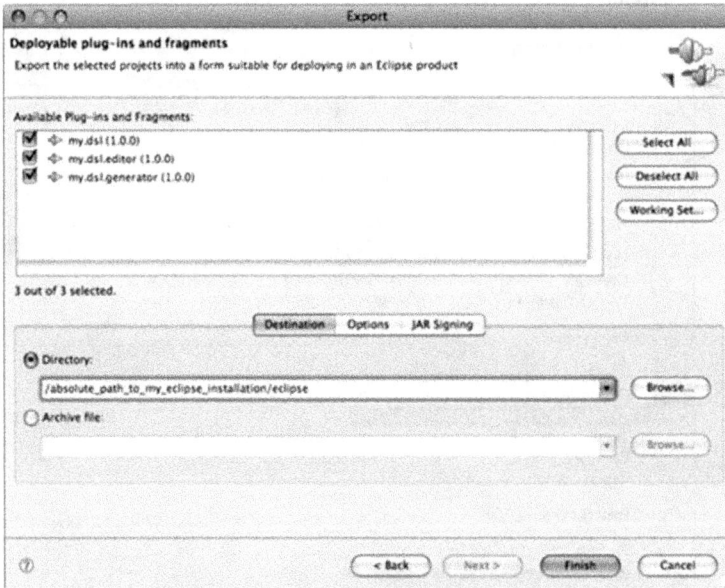

Figure 7.11. Deployment of the DSL plug-ins.

Figure 7.12. *Xpand* template.

The text analysis is divided into two separate tasks: the *lexing* and the *parsing*.

The *lexer* is responsible for creating a sequence of tokens from a character stream. Such tokens are identifiers, keywords, whitespace, comments, operators, and so on. *xText* comes with a set of built-in rules that can be extended or overwritten if necessary.

The *parser* gets the stream of tokens and creates a parse tree out of them.

Type Rules

The name of the rule is used as name for the metatype generated by *Xtext*.

Assignment tokens/Properties

Each assignment token within an *xText* grammar is not only used to create a corresponding assignment action in the parser but also to compute the properties of the current metatype. Properties can refer to the simple types such as String, Boolean, or Integer, as well as to other complex metatypes. It depends on the assignment operator and the type of the token on the right, that is, the actual type. There are three different assignment operators:

- Standard assignment "=": The type will be computed from the token on the right.
- Boolean assignment "?=": The type will be Boolean.
- Add assignment "+=": The type will be List. The inner type of the list depends on the type returned by the token on the right.

An example in FFMM project of these assignment operators is shown in Figure 7.13.

Cross References

Parsers construct parse trees not graphs. In the model, crosslinks are usually implemented by linking. However, *xText* supports the specification of linking information in the grammar, so that the meta-model contains cross references and the generated linker links the model elements automatically. Linking semantic can be arbitrarily complex. *xText* generates a default semantic which can be selectively overwritten.

Let's take a look at the optional extends clause. The rule name *Entity* on the right is surrounded by squared parenthesis (Figure 7.14). By default, the parser expects an identifier to point to the referred element.

Metatype Inheritance

After the definition of metatypes and their features, type hierarchies using the grammar language of *xText* have to be described. Different kinds of

```
DeviceSetting:
    cmds=CommandStatement    | sets=SettingStatement | gets=GettingStatement | uses=Use_Statement
    cppCode=CppCode          | ast=AssignStatement;

MTask:
"BEGIN MTASK" mtaskName=ID (mtaskDesc=STRING)? ";"
    (taskDeclarations+=Declarations)*
    (taskAction+=GenericStatement)*
"END MTASK";

GenericStatement:
    cmds=CommandStatement    | sets=SettingStatement | gets=GettingStatement | uses=Use_Statement
    cpp=CppCode              | ast=AssignStatement   |
    fst=ForStatement         | wst=WhileStatement    | ist=IfStatement    | ust=Util_Statement;

Util_Statement:
print=Print_ | delay=Delay_ | trigEvent=TrigEvent_;
```

Figure 7.13. Example of assignment operators in FFMM project.

"Features" (Figure 7.14) can be created by means of an abstract type rule such as shown in Figure 7.15.

The transformation creating the meta-model normalizes the type hierarchy automatically. This means that properties defined in all subtypes will be moved automatically to the common supertype.

The ID Token

The identifier token (ID) is the token rule expressed in AntLR grammar syntax, such as shown in Figure 7.16. The return value of the ID token is a String. Thus, if the usual assignment operator "=" is used, the assigned value will be of type String.

```
Entity :
    "entity" name=ID ("extends" superType=[Entity])?
    *{*
        (features+=Feature)*
    *}*;
```

Figure 7.14. Entity.

```
Script:
"BEGIN_SCRIPT" scriptName=ID ":"
    (scriptDeclarations+=Declarations)*
    (scriptAssignements+=ScriptAssignement)*
    (scriptDeviceDefinitions+=Definition_Statement)*
    (scriptDeviceConfigurations+=ConfigurationStatement)*
    (scriptDeviceSettings+=DeviceSetting)*
    (mtasks+=MTask)*
    (taskExecutionStatements+=TaskExecutionStatement)*
    // (rts=RunTaskSequence_)
"END_SCRIPT";

ScriptAssignement:
    (ast=AssignStatement | cpp=CppCode);

DeviceSetting:
    cmds=CommandStatement    | sets=SettingStatement | gets=GettingStatement | uses=Use_Statement |
    cppCode=CppCode          | ast=AssignStatement;

MTask:
"BEGIN_MTASK" mtaskName=ID (mtaskDesc=STRING)? ":"
    (taskDeclarations+=Declarations)*
    (taskAction+=GenericStatement)*
"END_MTASK";

GenericStatement:
    cmds=CommandStatement    | sets=SettingStatement | gets=GettingStatement | uses=Use_Statement|
    cpp=CppCode              | ast=AssignStatement   |
    fst=ForStatement         | wst=WhileStatement    | ist=IfStatement    | ust=Util_Statement;

Util_Statement:
print=Print_ | delay=Delay_ | trigEvent=TrigEvent_;

TrigEvent_:
"TRIG_EVENT" eventName=ID ";";

TaskExecutionStatement:
at=AddTask_ | atat=AddTaskAfterTask_ | atae=AddTaskAfterEvent_;
```

Figure 7.15. *Abstract type* rule.

Comments

There are two different kinds of comments automatically available (Figure 7.17) in any *xText* language. Note that those comments are ignored by the language parser by default.

Defining the MDSL

The goal was to create a simple scripting language for the test engineer; this problem has been addressed through the definition of a DSL. The test engineer has to follow the steps shown in Figure 7.18 to define, set, and execute a measurement task.

More precisely, the test engineer should first define the object (or device) that he intends to use, then configure its setting, and use it through appropriate commands, defined in device interfaces, which should be known by the test engineer. To make this task easier, the MDSL project provides one of the most useful things: the assistance to the measurement procedure definition.

While he writes the script, the test engineer can click on *CTRL+SPACE* to see the menu where all the possibilities are shown (Figure 7.19).

It is possible to appreciate the ease of writing and the flexibility of software. In the following, for the sake of comparison, two script fragments are shown. Figure 7.20 refers to a C++ script for permeability measurements, while Figure 7.21 shows the same procedure written in DSL. The improvements in clarity and conciseness are evident.

```
//LVPowerSupply Definition
Def_LVPowerSupply_:
    "LV_POWER_SUPPLY:" name=ID  ("WITH""("mod=Param","type_num=Param","ser_num=Param","pow_class=Param")")?;
    //static LVPowerSupply* createDevice( std::string name);
    //static LVPowerSupply* createDevice( std::string name, std::string type_num, std::string ser_num, std::string

//DAQmx Using
Use_LVPowerSupply_:
    "LV_POWER_SUPPLY:" name=ID;


//LVPowerSupply Configuration
Cfg_LVPowerSupply_:
    "LV_POWER_SUPPLY:" name=ID"WITH""("interfacenum=Param","devaddr=Param","timeout=Param")";
    //  void configure( int interfacenum, int devaddr, int timeout );
```

Figure 7.16. *Token* rule expressed.

```
// single-line comments and

/*
  mutli-line comments
*/
```

Figure 7.17. Comments.

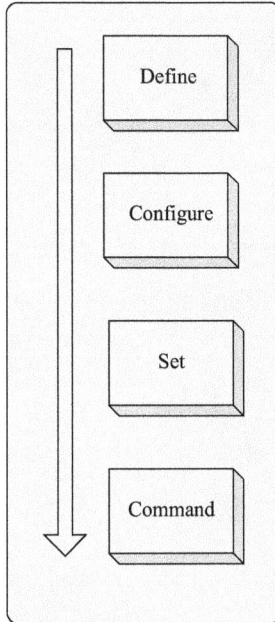

Figure 7.18. DSL test engineer steps.

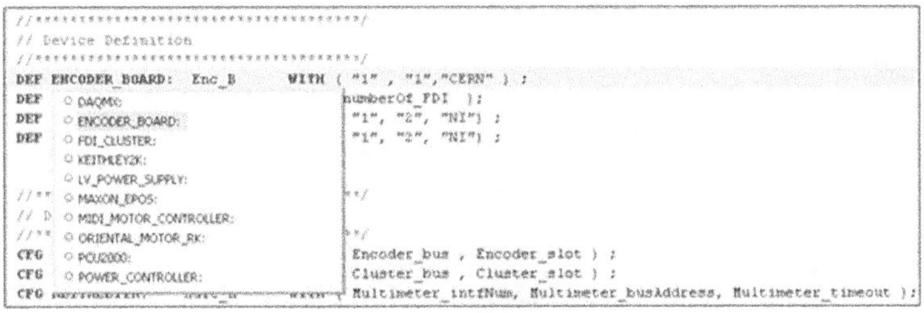

Figure 7.19. Assistance to the measurement procedure.

7.6   SOFTWARE QUALITY ASSESSMENT

7.6.1   ISO 9126 CHARACTERIZATION

The analysis of the software quality of FFMM was carried out by means of the tool Understand C++ [11]. Heuristic thresholds were employed, as proposed in literature [12–15], in order to define the metrics target values. An example of the metrics and the corresponding target values is reported in Table 7.2 [12, 16].

In particular, the complexity metrics (such as Essential Complexity and Cyclomatic Complexity [17]) measure the logic complexity of the software modules and hence the effort required for testing and maintaining

```cpp
#include <ffmm.h>
#include <sstream>
#include <math.h>
using namespace ffmm::core::events;
using namespace ffmm::core::devices;
using namespace std;
_FFMM_INITIALIZE
#include "core/devices/FdiCluster.h"
#include "core/devices/EncoderBoard.h"
#include "core/devices/Keithley2k.h"
#include "core/devices/DAQmx.h"
#include "core/utils/FdiClusterDataConversion_byn2Ascii.h"
#include "core/events/IFdiClusterListener.h"

std::string Cluster="FDI_Cluster_1";
std::string Encoder="Encoder_Board";
std::string Multimeter="Keithley2k";
std::string DAQM="NI_DAQ";
const int Encoder_slot=13;
const int Encoder_bus=4;
const int Encoder_Channel=1;
const int Encoder_mode=1;
const double Encoder_freq=2048;
const int Multimeter_intfNum=0;
const int Multimeter_busAddress=16;
const int Multimeter_timeout=100;
const int numberOf_FDI = 3;
const int surceStop = 1;
int Cluster_slot[numberOf_FDI]={12,11,10};
int Cluster_bus[numberOf_FDI]={4,4,4};

double Cluster_abs_gain_ = 1.0;
double Cluster_comp_gain_ = 10;
int SamplePerTurn = 1024;
int numberOfTurn = 4;
U32 AcquisitionBufferSize;
std::string Daq_channel_name = "AO_Ch";
std::string Daq_task_name = "Trap_G";
int Daq_channel = 0;
int Daq_timeOut = 200;
int Daq_generatioMode = 0;
const double Daq_sample_rate = 1000;
int Daq_minVolt = -10;
int Daq_maxVolt = 10;
std::string path_name;
double epsC = 0.1;
int measurementCycle;
double    plateaux[38]   =   {0,-0.1,0.1,-0.2,0.2,-0.3,0.3,-0.4,0.4,-0.5,0.5,-0.6,0.6,-0.7,0.7,-0.8,0.8,-0.9,0.9,-1,1,-
1.2,1.2,-1.4,1.4,-1.6,1.6,-1.8,1.8,-2,2,-3,3,-5,5,-10,10,0};
BEGINSCRIPT
        NI_Daq->setTimingTrigger(Daq_sample_rate, 0, numOfSamples);
        NI_Daq->startVoltage(signal, numOfSamples);
        NI_Daq->waitGeneration();
        NI_Daq->setTimingTrigger(Daq_sample_rate, 0, numOfSamples);
        NI_Daq->startVoltage(signal, numOfSamples);
        NI_Daq->waitGeneration();
        Poco::DynamicAny plat;
        while(!demagnetized)
        {
                Poco::DynamicAny plat(plateau*4);
                environment->console->writeln(plat.convert<std::string>());
                if (plateau >= value1)
                {
                        old_plateau = plateau;
                        plateau = plateau/1.5;
                }
```

Figure 7.20. The part of the Script in C++.

```
            else if (plateau >= value2)
            {
                    old_plateau = plateau;
                    plateau = plateau/1.2;
            }
            else
            {
                    old_plateau = plateau;
                    plateau = plateau/1.1;
            }

            signal      =      NI_Daq->createRampOrSquareVoltage(old_plateau,1/Daq_sample_rate,-
plateau,timePlateau, 1.5,0,Daq_sample_rate,&numOfSamples);
            NI_Daq->setTimingTrigger(Daq_sample_rate, 0, numOfSamples);
            NI_Daq->startVoltage(signal, numOfSamples);
            NI_Daq->waitGeneration();

            signal = NI_Daq->createRampOrSquareVoltage(-

plateau,1/Daq_sample_rate,plateau,timePlateau, 1.5,0,Daq_sample_rate,&numOfSamples);
NI_Daq->setTimingTrigger(Daq_sample_rate, 0, numOfSamples);
            NI_Daq->startVoltage(signal, numOfSamples);
            NI_Daq->waitGeneration();

            if (plateau <= 0.001)
            {
                    demagnetized = 1;
            }
    }
```

Figure 7.20. (*Continued*).

```
DEF ENCODER_BOARD:  Enc_B       WITH ( "1" , "1","CERN" ) ;
DEF FDI_CLUSTER:    Cluster_1   WITH (numberOf_FDI  );
DEF KEITHLEY2K:     Mult_M      WITH ( "1", "2", "NI") ;
DEF DAQMX:          NI_Daq      WITH ( "1", "2", "NI") ;


//*************************************/
// Device Configuration
//*************************************/
CFG ENCODER_BOARD: Enc_B        WITH ( Encoder_bus , Encoder_slot ) ;
CFG FDI_CLUSTER:   Cluster_1    WITH ( Cluster_bus , Cluster_slot ) ;
CFG KEITHLEY2K:    Mult_M       WITH ( Multimeter_intfNum, Multimeter_busAddress, Multimeter_timeout );
CFG DAQMX:         NI_Daq       WITH ( Daq_channel_name, Daq_task_name, Daq_channel,  Daq_timeOut, Daq_generatioMode  );

//*************************************/
// Device Setting
//*************************************/
CMD FDI_CLUSTER: Reset ( Cluster_1, 0);
CMD FDI_CLUSTER: Reset ( Cluster_1, 1);
SET FDI_CLUSTER: Params2 ( Cluster_1, spt1, SamplePerTurn, Cluster_abs_gain_, Cluster_comp_gain_, CONT, 500000, spt2,10);
SET FDI_CLUSTER: Stop_Source ( Cluster_1, surceStop);
SET DAQMX:   VoltageRangeOutputChannel ( NI_Daq, Daq_minVolt, Daq_maxVolt);
CMD FDI_CLUSTER: Calibrate_Gain (Cluster_1, 0, 1.0);
CMD FDI_CLUSTER: Calibrate_Gain (Cluster_1, 1, 1.0);
```

Figure 7.21. The same Script of Figure 7.20 in DSL.

them. The Object-Oriented metrics, taken from well-known metrics suites [18–21] (LCOM, FAN IN, CBO, RFC, WMC, and DEPTH) measure the extent to which features typical of Object-Oriented systems are exploited (e.g., inheritance) or achieved (e.g., lack of coupling and cohesion). Table 7.3 reports a short summary of size metrics computed on FFMM.

At first glance, the complexity metrics (Essential Complexity and Cyclomatic Complexity) show that, although in FFMM the average complexities respect the heuristic upper bounds, the maximum values exceed them in a significant way (Table 7.4). This means that the complexity is

Table 7.2. Complexity and Object-Oriented metrics with their target values

Metric	Target
Cyclomatic Complexity (CYCLO)	≤ 10
Essential Complexity (ESS)	≤ 4
Class Depending Child (CDC)	*FALSE*
Class Depth (DEPTH)	≤ 7
Multiple Inheritance (FAN IN)	≤ 1
Response for Class (RFC)	$\leq (WMC * DEPTH) + 1$
Coupling between Objects (CBO)	≤ 2
Lack of Cohesion of Methods (LOCM/LCOM)	≥ 0.75
Weighted Methods for Class (WMC)	≤ 14

Table 7.3. FFMM 3.0 size metrics summary

Blank Lines	4'115
Classes	96
Code Lines (LOC)	16'253
Comment Lines	6'977
Comment to Code Ratio	0.43
Declarative Statements	4'779
Executable Statements	8'642
Files	131
Functions	1'082
Inactive Lines	172
Lines	28'119

concentrated in few points that need to be simplified in order to decrease the effort required for software testing and maintenance. Analogous remarks can be made from the analysis of the Object-Oriented metrics (Table 7.5): Most of them show maximum values significantly exceeding the heuristic thresholds, potentially causing problems to system developers and users.

The values of the FAN IN metric exceeding the threshold are the result of a conscious design choice, since all the devices implemented in FFMM inherit from two abstract classes. These two classes are completely independent from each other, therefore the multiple inheritance is not expected to cause any undesired side effects.

Table 7.4. FFMM 3.0 complexity metrics

Metric	Average	Max	Std	Target	% OK (program units)
ESS	1.2	19	1.1	≤ 4	99
CYCLO	2.4	40	3.2	≤ 10	97

Table 7.5. FFMM 3.0 Object-Oriented metrics

Metric	Average	Max	Std	Target	% OK (classes)
CDC	FALSE	FALSE	-	FALSE	100
DEPTH	1	4	1.2	≤ 7	100
FAN IN	0.7	2	0.7	≤ 1	85
RFC	20	138	31	$\leq (WMC$ *DEPTH$)+1$	45
CBO	3.9	21	4.6	≤ 2	50
LOCM/LCOM	0.42	1	0.40	≥ 0.75	40
WMC	13	103	18	≤ 14	74

An attempt to use internal metrics, computed on the source code, according to the model ISO 9126 for an automatic metric based quality control, is presented in [22, 23]. This approach proposes a set of metrics and a quality matrix mapping them into factors and criteria of the model.

Heuristic thresholds [12–15] were used to evaluate the percentages of classes exceeding the acceptable values for the considered metrics. The quality matrix allows the translation of these properties into the levels at which quality characteristics are achieved.

The quality model presented in [23] refers to an adapted quality model where reusability replaces usability, evaluated on the basis of both non-object-oriented metrics (mainly referring to size) and to Object-Oriented metrics (mainly referring to cohesion, coupling, and inheritance). The results obtained through this approach are shown in Figures 7.22 and 7.23. The values vary in the range [0,1], with 0 corresponding to the best quality level. The aforementioned connections between static metric analysis and internal software qualities have been proposed recently, are only partially validated, and still need improvements and confirmations to be widely accepted by the technical community. In other words, the automatic assessment of software quality defined by the ISO standards using internal metrics is still an open question. Therefore, the results presented in this section are intended to be only as suggestive of the quality level achieved

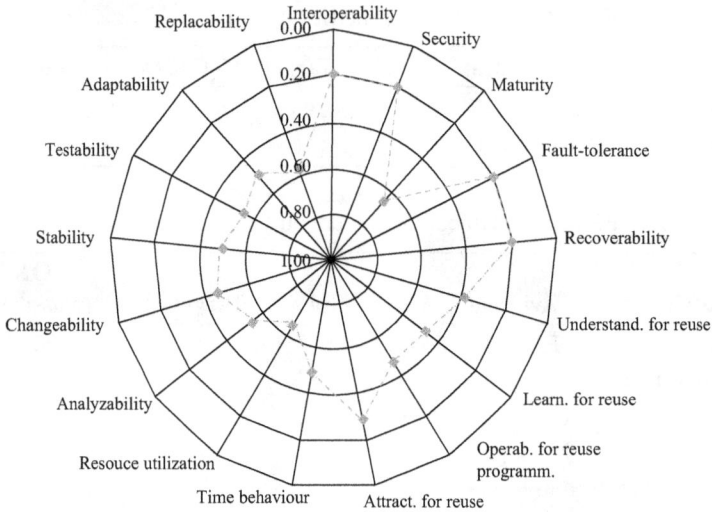

Figure 7.22. ISO 9126 subcharacteristics in FFMM 3.0 (0 indicates the best quality level).

by FFMM. Furthermore, no direct indications are provided on how to improve quality. For this reason, in the following, the quality analysis is complemented by introducing a more practical approach.

7.6.2 QUALITY PYRAMID CHARACTERIZATION

The Overview Pyramid obtained from the FFMM source code by means of the analysis tool inFusion [24], is presented in Figure 7.24.

In order to provide a graphical aid to its interpretation, different grey tonalities are associated to the computed proportions. In particular, a medium grey rectangle is used if the value contained in it is closer to the low threshold, a dark grey rectangle if it is closer to the average threshold, and a light grey rectangle if it is closer to the high one. An analysis of the resulting computed proportions allows the following conclusions to be drawn for the FFMM source code:

- Class hierarchies tend to be tall and narrow, that is, inheritance trees tend to have many depth-levels and base-classes with few directly derived subclasses.
- Classes tend to be rather large (they define many methods) and organized in fine-grained packages (few classes per package).
- Methods tend to be average in length and have an average logical complexity (conditional branches).

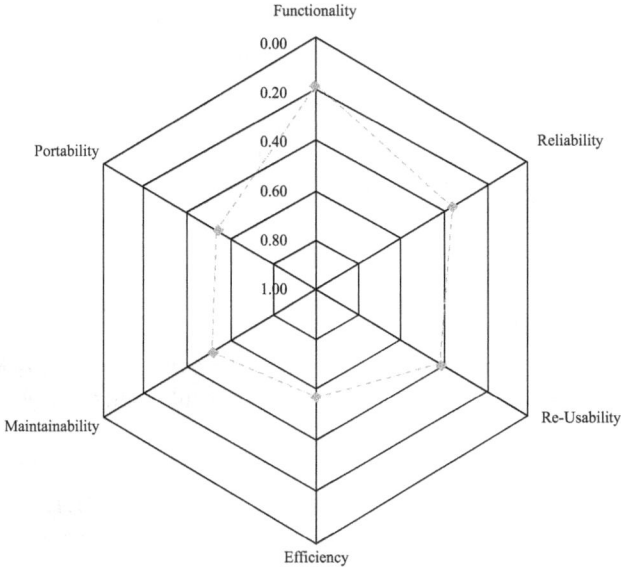

Figure 7.23. ISO 9126 characteristics in FFMM 3.0 (0 indicates the best quality level).

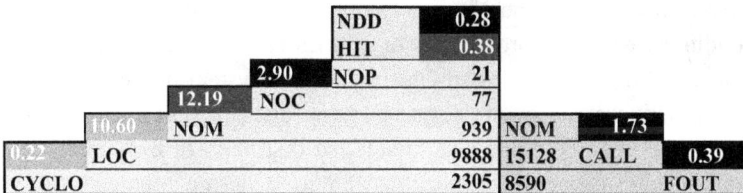

Figure 7.24. Overview Pyramid for the FFMM 3.0 source code.

In the development of a software system it is often difficult to find the appropriate design, the responsibilities of objects, and their distribution. Once this has been done, it is important to be able to state whether the complexity due to the design choices is balanced by the benefits they introduce.

Although the Overview Pyramid allows the graphic characterization of an Object-Oriented system through the quantification of some suitable chosen metrics, it is not enough to completely understand and evaluate the design. Metrics and thresholds must be meaningful and put in the right context [12, 25], in order to assess the quality level of the project design and eventually to ameliorate it. An application, a class, and a method should be implemented in a harmonious way, in terms of size, complexity, and functionality, with respect to itself, its collaborators, and its ancestors and descendants. In other words, the system has to achieve an overall harmony, composed of three distinct measurable parts [12]:

- **Identity harmony**, related to the extent to which a software entity implements a specific concept and how well (Does it implement too many things? Is it not doing enough to exist as an autonomous entity?).
- **Collaboration harmony**, expressing the extent to which an entity collaborates with others, and how well (Does an entity use other entities? How many?).
- **Classification harmony**, combining elements of the other two harmonies in the context of inheritance (Are inherited services used by subclasses? Some or all of them?).

The design evaluation, from the point of view of the aforementioned harmonies, intents ,therefore, on questioning if every software entity has appropriate place, size, and complexity within the system.

Disharmonies are revealed by means of metric-based heuristics to detect and locate Object-Oriented design flaws from the source code. For the quantification of complex design rules, the evaluation of a single metric is not sufficient. Therefore, detection strategies, composed of logical conditions based on proper sets of metrics and thresholds, are used to evaluate design quality of an Object-Oriented system through quantifying deviations from good design heuristics and principles. Design rules can in this way be made quantifiable, so that parts of the source code with specific properties (typically denoting a design problem) can be detected. The result of this stage of detection is a list of software entities suspected of being affected by some flaw. These entities need subsequently to be inspected to find those that cause the most severe problems and determine how to refactor them. This insight into the class structure is needed to understand its static structure, that is, the way attributes are accessed, methods called, inheritance used, and decide if there is need for intervention.

Among the 11 design disharmonies classified in [12], the three categories presented earlier, those that were detected in FFMM, are discussed in the following. It is worth pointing out that no significant code duplication was found. This is an important achievement of the FFMM design, because code duplication can be very harmful: It breaks the uniqueness of certain functionalities within the entities in the system, and causes an increase of size, complexity, error proneness, and also problems of coevolution of duplicated code.

Identity Disharmonies

A design flaw is referred to as an identity disharmony if it affects single entities, such as classes and their methods. The disharmonies of a single entity are related to three aspects: size, interface, and implementation. These aspects can be summarized by three rules of identity harmony [12]:

- A harmonious size is a desirable feature of operations and classes.
- The interface of each class should be a set of services, which implement only one responsibility and provide a unique behavior.
- Data and operations semantically belonging to the same class should collaborate harmoniously within that class.

A violation in the proportion rule, often due to excessive size and complexity of a class, is a typical and easily recognizable indication of identity disharmonies, and one of the causes could be a massive presence of code duplication. Another sign of disharmony could be a reduced cohesiveness of behavior (presentation and implementation rule) and the tendency to include in a single class too many features and services, thus producing a God Class [26, 27]. A consequence of this disharmony is usually a Feature Envy [27], because if a class becomes a God Class, other classes tend to become Data Classes [26], which are simple data containers not providing much functionality. The God Class will contain methods which will be likely to access attributes of other classes rather than those of the class itself. The Feature Envy consists of this excessive interest of a class in external data.

God Class

The God Class design flaw is due to classes that tend to centralize too much intelligence and functionalities of the system [26]. A God Class tries to do too much work, collaborating with a set of simple classes to which it delegates only trivial tasks, and using extensively data contained in other classes. This situation has often a negative effect on reusability and understandability.

In FFMM, the three classes were detected as God Classes because: (a) some of their methods access directly (or via getter and setters) attributes from external classes, (b) the methods are very complex, that is, have many branches, and (c) the classes are noncohesive with respect to the way methods use the attributes of the class. An analysis of the results shows that:

1. The methods accessing attributes from external classes are the same for the three God Classes, and are used for data type conversion, bus configurations, or thread handling. In all cases, they result from precise implementation choices and are not considered a source of problems.
2. The cohesiveness of all the classes is only slightly below the limit of one third. As a consequence of these considerations, the problem represented by the God Classes can be classified as noncritical. Anyway, it could be useful to reduce the complexity of the methods, because this might affect understandability, usability, and maintainability.

Data Class

A Data Class [27, 28] is a trivial class acting as simple data holder, not providing complex functions, but rather extensively used by other classes. The lack of functions may indicate that the data contained in the class are not kept in the same place with the related behavior. In other words, Data Classes provide almost no function through their interfaces, which mainly exposes data fields either directly or through accessory methods.

In FFMM, the three detected Data Classes do not produce any Feature Envy. This implies that, although they do not provide complex functionality, the data they contain are not significantly accessed by methods of other classes. The problem represented by the detected Data Classes can be therefore reappraised if correctly put into this perspective. They do not contain misplaced data needed by other parts of the system, but are the result of a precise design choice for handling the communication buses and their configurators.

Brain Class

This design flaw refers to the situation where an excessive amount of intelligence is accumulated in a single class, often concentrated in Brain Methods. It recalls the God Class, but the two disharmonies are distinct. A God Class, besides being complex and centralizing a large amount of the system intelligence, breaks the encapsulation principle by accessing attributes from other classes, and showing a lack of cohesion. The Brain Class detection strategy is complementary to that of the God Class, catching very complex classes which do not break encapsulation and do not manifest a significant lack of cohesion.

The main characteristic of a Brain Class is that it is very likely to contain Brain Methods; therefore, the first improvement action should concern these methods, as discussed in the relative section. On the other hand, if the class is detected as a Brain Class owing to its lack of cohesion, it should be split into an appropriate number of more cohesive classes. However, it is often the case where a Brain Class does not cause any significant problem in the system, that is, if it is simply a mature complex utility class. In such cases, if no maintenance problems arise during the system history [29], it is not worth starting a costly refactoring phase just to get better metric values.

In FFMM, the six Brain Classes contain only 1 or 2 Brain Methods, and are characterized by high complexity and low cohesion. As corrective actions, the Brain Methods should first be split. In case of maintenance problems, the involved classes could also be split into smaller and more cohesive units.

Feature Envy

One of the goals of Object-Oriented Programming is to keep together a set of data and the operations processing that data. The Feature Envy design flaw [27] refers to methods that access, directly or through accessory methods, more data of other classes than of their own class. The Feature Envy disharmony is often an indication that a method is misplaced and should be moved to another place in the system. The problem can be solved if the method, or a part of it, is extracted from its current class and moved to the class containing the envied data [27, 28].

The Feature Envy problem is often due to the presence of Data Classes, which make the classes using them to envy their data. When a method is affected by Feature Envy, it is probable that there are data classes among the classes from which the method accesses data. The fact is that in FFMM the two methods affected by Feature Envy do not access Data Classes, highlighting that the problem is isolated to two methods which are used in handling the polling tasks of a class. Even though this is detected as a violation of the encapsulation principle of the Object-Oriented design, it is the result of an implementation choice for the threads executing the polling methods. Moreover some metrics are very close to the lower detection threshold. It is therefore concluded that the observed Feature Envy does not represent a critical issue in the system, and that the cost of a possible code improvement action would not be balanced by proportionate benefits.

Brain Method

Brain Methods centralize excessively the functionality of a class (proportion and implementation rules), in a way similar to that of a God Class which centralizes the functionality of the whole system or of one of its parts, making it hard to understand, maintain, and reuse [12].

The 14 Brain Methods detected result in being rather complex (both for conditional branching and nesting) and long, and they employ a very large number of variables. For the negative impact of this design flaw on understandability, maintainability, and reusability, these methods require a refactoring. In this case, the Brain Methods do no not exhibit either significant code duplication or Feature Envy, thus there is no cloned code to remove or Data Classes to where some of the behavior complexity can be moved. In literature it is suggested that in almost all cases a Brain Method should be split into one or more simpler methods [27], by finding appropriate cutting points.

Collaboration Disharmonies

Collaboration disharmonies are design flaws that affect several entities at once in terms of the way they collaborate to perform a specific functionality. All the authors propose low coupling as a design rule for Object-Oriented system.[5] Anyway, a tradeoff has to be found between the intent of low coupling and the need for a certain amount of collaboration among objects of the same system. The collaboration harmony consists in the achievement of a balance between the aforementioned opposing demands.

All this can be summarized in the following rule:

- Collaborations should be only in terms of methods invocations and should have a limited extent, intensity, and dispersion;

where extent refers to the number of other classes; intensity, to the number of services provided by other classes; and dispersion, to the distance of collaborating classes (two classes can be in the same hierarchy, package, etc.). The rule refers both to incoming and outgoing dependencies. Excessive outgoing dependencies are undesirable because they make a class

---

[5]To this purpose, an event handling infrastructure (Chapter 5) is used whenever possible in FFMM, minimizing the coupling introduced by the necessary communications among objects.

more vulnerable to changes and bugs from other classes. On the other hand, excessive incoming dependencies are undesirable because they create the need of stability and, therefore, make the class less evolvable.[6] Collaboration disharmonies are captured using two detection strategies: Intensive Coupling and Dispersed Coupling. The former refers to the case where a method uses intensively a reduced number of classes, the latter to situations where the dependencies are dispersed over many classes. Moreover, on the server method side, it might happen that a method is excessively invoked by numerous methods located in various other classes (Shotgun Surgery [27]). In this case, a small change in a part of the system can cause a lot of changes in many other classes.

Intensive Coupling

Coupling reduction has the main aim of using a component apart from others, or making easy its replacement with another one. A usual refactoring action that allows solving the problem consists of defining a new more complex service in the provider class, and replacing the multiple calls with a single call to the new method. In some cases the design flaw might be due to misplaced operations. This is in the case of FFMM, where the intensive coupling involves two methods contained in the classes modeling two motor controllers. These methods access many (11) methods of another class. The origin of the problem is the fact that the methods were developed at a very low abstraction level in the system, that is, in a device. In fact, each method involves two devices, and once inserted in one of them, it has to call many methods of the other class. On the basis of this consideration, it is clear that it should be implemented at a higher level, as a measurement routine employing both the aforementioned devices. Placing the method into a library of measurements routines would solve the coupling problem, besides removing the code duplication, by leaving only one instance of the method.

Shotgun Surgery

Unlike the intensive coupling, the Shotgun Surgery refers to the cases where the incoming dependencies can cause problems, that is, where a change in

---

[6]It is to be noted that, at the same time, incoming dependencies could mean high degree of code reuse in the system, with the condition of having stable interfaces.

an operation implies many changes to a lot of other components (methods and classes) in the system [27]. Similarly, if this method has bugs, it will have a significant negative effect on the parts of the system that are using it. In such a situation, maintenance and evolution problems could arise.

A possible refactoring action to solve the problem consists in moving more responsibility to the classes containing Shotgun Surgeries, above all, for small and noncomplex methods and classes with tendency to become Data Classes [12]. Anyway, this is not the case in FFMM, where this design flaw involves five methods of two classes designed in order to favor code reuse for interacting with the user and handling the communication buses (and, consequently, massively accessed by methods in other classes). The characteristics of these methods come therefore from a conscious design choice, and do not cause any problems under the assumption that they are used through stable interfaces.

REFERENCES

[1] Jain, P., and D.C. Schmidt. June 1997. "Service Configurator." *Proceedings of the Third USENIX Conference on Object-Oriented Technologies and Systems.* Portland, OR: Usenix.

[2] Arpaia, P., and M.L. Bernardi. 2007. "Executive Project." FFMM Project Tecnichal Notes, CERN -07-11. Giuseppe Di Lucca University of Sannio.

[3] Lavender, R.G., and D.C. Schmidt. September 1995. "Active Object—An Object Behavioral Pattern for Concurrent Programming." *Proceedings of the Second Pattern Languages of Programs Conference (PLoP).* Monticello, IL: Addison-Wesley Longman.

[4] Sommerville, I. *Software Engineering.* 1992. 4th ed. Harlow, England: Addison Wesley.

[5] Arpaia, P., M. Buzio, L. Fiscarelli, and V. Inglese. November 2012. "A Software Framework for Developing Measurement Applications Under Variable Requirements." *AIP Review of Scientific Instruments* 83, no. 11. doi: 10.1063/1.4764664

[6] Arpaia, P., M.L. Bernardi, G. Di Lucca, V. Inglese, and G. Spiezia. September 20–21, 2007. "Fault Self-Detection of Automatic Testing Systems by Means of Aspect Oriented Programming." *IMEKO '07 Conference.* Iasi, Romania: IMEKO.org.

[7] Ballarino, A., G. Montenero, and P. Arpaia. 2014. "Transformer-based Measurement of Critical Currents in Superconducting Cables: Tutorial 51." *Instrumentation & Measurement Magazine, IEEE* 17, no. 1, pp. 45–55. doi: 10.1109/MIM.2014.6782997

[8] Ramsbottom, H.D., S. Ali, and D.P. Hampshire. 1996. "Response of a New Ceramic-Oxynitride (Cernox) Resistance Temperature Sensor in High

Magnetic Fields." *Cryogenics* 36, no. 1, pp. 61–63., doi: 10.1016/0011 -2275(96)80772-7

[9] Midi-Inginierie. 2008. *Midi Motor Controller Simpa*, http://www.midi-inginerie.com/NEW/DataSheet/__SIMPA/sm01_cd_fr.pdf

[10] Agilent Technologies. 2013. "PXI Interoperability—How to Achieve Multi-Vendor Interoperability in PXI Systems." *Application note*. Santa Clara, CA: Agilent technologies.

[11] Understand C++. 2014. http://www.scitools.com/products/understand/

[12] Lanza, M., and R. Marinescu. 2006. Object-Oriented Metrics in Practice. Secaucus, NJ: Springer-Verlag New York, Inc.

[13] French, V.A. September 1999. "Establishing Software Metric Thresholds." *In WSM 99: International Workshop on Software Measurement*, pp. 43–50. Lac Supérieur, QC: IEEE.

[14] Le metriche e il loro utilizzo nello sviluppo del software (in Italian). 2005. http://www.dia.uniroma3.it/ torlone/sistelab/annipassati/sbavaglia.pdf

[15] Nikora, A.P. et al. 2002. "Software metrics in use at JPL applications and research." http://trs-new.jpl.nasa.gov/dspace/bitstream/2014/8671/1/02-1209.pdf

[16] Metrics available in understand C++. 2014. http://www.scitools.com/ documents/metrics.php

[17] McCabe, T.J. December 1976. "A Complexity Measure." *IEEE Transactions on Software Engineering* 2, no. 4, pp. 308–20. doi: 10.1109/ TSE.1976.233837

[18] Chidamber, S.R., and C.F. Kemerer. June 1994. "A Metrics Suite for Object Oriented Design." *IEEE Transactions on Software Engineering* 20, no. 6, pp. 476–93. doi: 10.1109/32.295895

[19] Li, W., and S. Henry. May 1993. "Maintenance Metrics for the Object Oriented Paradigm." In *IEEE Proceedings of the First International Software Metrics Symposium*, pp. 52–60. Baltimore, MD: IEEE.

[20] Bieman, J.M., and B.K. Kang. April 1995. "Cohesion and Reuse in an Object-Oriented System." In *Proceedings of the ACM Symposium on Software Reusability, ACM*, vol. 20, pp. 259–62. Seattle, WA, USA: ACM.

[21] Hitz, M., and B. Montazeri. October 1995. "Measure Coupling and Cohesion In Object-Oriented Systems." In *ISACC'95: Proceedings of International Symposium on Applied Corporate Computing*, pp. 24, 25, 274, 279. Monterrey, Mexico: Citeseer.

[22] Lincke, R., and W. Lowe. July 2006. "Validation of a Standard- and Metric-Based Software Quality Model." In *10th ECOOP Workshop on Quantitative Approaches in Object-Oriented Software Engineering (QAOOSE)*. Nantes, France: Växjö University.

[23] Lincke, R. and W. Lowe. April 2007. "Compendium of Software Quality Standards and Metrics." http://www.arisa.se/compendium/

[24] Invensys. 2010. Infusion http://iom.invensys.com/EN/pdfLibrary/Datasheet_InFusion_InFusionView_08-10.pdf

[25] Salehie, M., S. Li, and L. Tahvildari. 2006. "A Metric-Based Heuristic Framework to Detect Object-Oriented Design Flaws." In *International Collegiate Programming Contest (ICPC) 2006: Proceedings of the 14th IEEE International Conference on Program Comprehension*, pp. 159–68. Athens Greece: IEEE.

[26] Riel, A. 1996. *Object-Oriented Design Heuristics*. Boston, MA: Addison Wesley.

[27] Fowler, M., K. Beck, J. Brant, W. Opdyke, and D. Roberts. 1999. Refactoring: Improving the Design of Existing Code. Boston: Addison Wesley

[28] Demeyer, S., S. Ducasse, and O. Nierstrasz. November 2013. "Object-Oriented Reengineering Patterns." Boston, MA: Morgan Kaufmann.

[29] Rajiu, D., S. Ducasse, T. Girba, and R. Marinescu. October 2004. "Using History Information to Improve Design Flaws Detection." In *CSMR'94: Proceedings Eighth Euromicro Working Conference on Software Maintenance and Reengineering*, pp. 223–32. Los Alamitos, CA: IEEE. Computer Society.

CHAPTER 8

FRAMEWORK COMPONENT VALIDATION

The test of the machine is the satisfaction it gives you. There isn't any
other test.
If the machine produces tranquility it's right.
If it disturbs you it's wrong until either the machine or your mind is
changed.
　　　　　—Robert M. Pirsig, *Zen and the Art of Motorcycle Maintenance:*
　　　　　　　　　　　　　　　　　　　　An Inquiry into Values

8.1　OVERVIEW

In this chapter, the on-field validation of the main framework components,
the *Fault Detector*, the *Synchronizer*, the Measurement Domain Specific
Language (MDSL), and the *Advanced Generator of User Interfaces* is
highlighted in the context of the Flexible Framework for Magnetic Measurement (FFMM), developed at CERN in cooperation with the University of Sannio. Each component is validated in a different case study of the
FFMM at CERN. The *Fault Detector* is analyzed in a rotating coil system
for superconducting magnet testing by assessing its (a) internal quality, in
terms of the modularity improvement derived from the Aspect-Oriented
Programming (AOP) introduction, and (b) external quality, by verifying
these benefits do not introduce side effects during run-time performance.
The *Synchronizer* is validated in a magnetic permeability measurement
system, used for validating the twofold expected advantages of the simplification of the measurement script and the speeding up of its definition.

The Domain Specific Language (DSL) is evaluated in a superconducting magnet testing and a magnetic permeability measurement in order to verify its benefits in terms of simplicity, effectiveness, and flexibility in the production of measurement software applications. Finally, the performance of the *Advanced Generator of User Interfaces* is validated in a magnetic permeability measurement system by featuring how test engineers are supported during usual FFMM operations at CERN.

8.2   FAULT DETECTOR

In this section, the validation of the AOP-based architecture of fault detection software components illustrated in Chapter 4 is presented. In particular, a case study on a rotating coil system for superconducting magnet testing is illustrated as a highlight: (a) the actual improvement of the software quality of the AOP version with respect to the previous existing OOP version for the fault detector, and (b) that the new AOP architecture does not have a negative impact on run-time performance of the overall system (due to a possible overhead of the aspect runtime interception). Thus, the software quality attribute of modularity is assessed to highlight the first point and some runs of the results of tests on an adequately instrumented version of the system are illustrated in order to verify the second point.

8.2.1   A CASE STUDY ON ROTATING COILS

FFMM has been employed at CERN for characterizing the superconducting magnets of the Large Hadron Collider (LHC). In particular, a huge measurement effort has been devoted to the investigation of the dynamic field errors ("*multipoles*") of the main dipole magnets [1]. Rapidly varying magnetic fields are measured by a new measurement station with high-speed rotating coils units and Fast Digital Integrators (FDI) [2].

The *Fault Detector* has been tailored on the FFMM, by specifying: device under test, quantity to be measured, measurement instruments, measurement circuit configuration, measurement algorithm, and data analysis. The objective was to handle anomalous working events in the actual measurement application [3].

Next, (a) the *measurement procedure*, (b) the *analysis of fault detection software*, (c) the *modularity comparison*, and (d) the *performance verification*, are illustrated.

8.2.2 THE MEASUREMENT PROCEDURE

The measurement procedure is based on the "rotating coil" method illustrated in Section 6.2.1 (Figure 8.1 [3]): A set of coil-based transducers are placed in the magnet bores, supported by a shaft turning coaxially inside the magnet.

The coil signal is integrated according to Faraday's law in the angular domain, by exploiting the pulses of an encoder mounted on the shaft, in order to get the induction field. Several coil segments are placed on the shaft by covering the length of the magnet. Each segment, in turn, is made up of three overlapped coils: The external one measures the mean field (absolute signal), while the series connection of the external and the central coils in opposition of phase allows the main field to be deleted in order to measure the field harmonics only (compensated signal). The *field quality* in the accelerator magnets is expressed in terms of the magnitude of undesirable harmonics (multipoles).

The coil shaft inside the magnet is turned by the *Micro-Rotating Unit* (μRU) whose motor is driven by a controller (Maxon Epos 24). The magnet under test is supplied power converters with digital control, with varying capacity depending on the test conditions: Tests are carried out in cold (up to 1.9 K) and warm (room temperature) conditions by using a 14 kA, 15 V, and a 20 A, 135 V power converter, respectively. The current is read by a digital multimeter through a high-accuracy Direct Current-Current Transformer (DCCT). The coil signals are integrated into the angular domain by digital integrators (i.e., an FDI [4], implemented by the *FastDI* class), by exploiting the trigger pulses coming out from a conditioning board (developed at CERN, implemented by the *EncoderBoard* class), suitably processing the output of the encoder mounted on the mRU. The FDI boards, the encoder conditioning board, and the motor controller are remotely controlled by a PC running the test program created by FFMM, produced according to a suitable script [5].

8.2.3 ANALYSIS OF FAULT DETECTION SOFTWARE [6]

The rotating coil testing technique is a typical application involving several different devices, each one with its own state and error messages. During the operation of the measurement station of Figure 8.1, one or more faults can affect the devices at different levels. At the lowest level, a fault can influence one of the communication buses (e.g., PXI, RS-232, WorldFIP, IEEE-488 GPIB). At this level, possible faults created on the

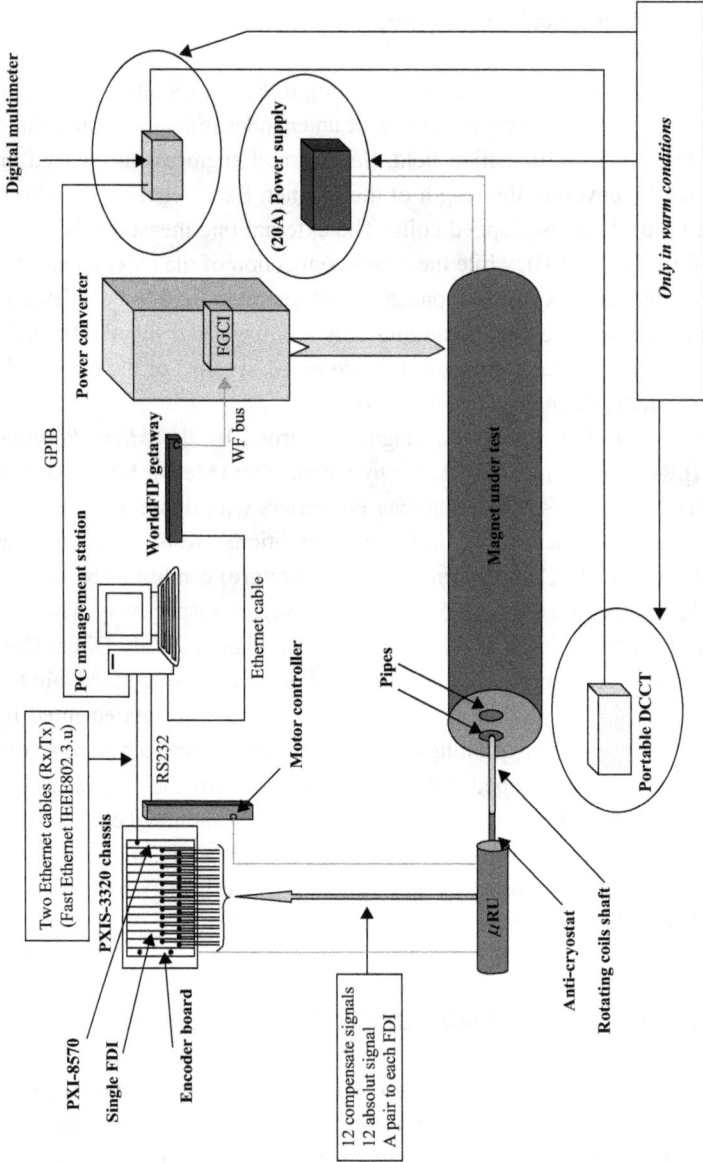

Figure 8.1. Layout of the rotating coil measurement setup [3].

buses are: (a) communication timeout, (b) device not found on the bus, and (c) error on an open, read, write, or close command. All these kinds of faults require that the data acquired until the fault occurrence be saved, log some of the diagnostic information about the state of the measurement station, and reset the devices in order to return them back to a consistent state.

At a higher level, some faults can involve the devices controlled through the communication buses, namely the FDI, the encoder board, and the motor controllers. In particular, the FDI can be affected by the following faults: (1) A timeout can occur during the time interval between the transmission and the execution of a command, or when the measurement starts and the integrator waits for the trigger pulses from the encoder; (2) an inconsistency of the internal status of the instrument; and (3) a wrong parameter value is set. The last two fault conditions can occur for the encoder board and the motor controllers. For the motor controller, another fault can arise from the handshake procedure on the communication bus (RS-232).

Within the AOP fault detection architecture, faults of the type (1) and (2) are detected by means of specific pointcut expressions associated to the advices capturing the access to an internal devices' status and eventually sending fault events to interested components. Conversely, the faults of type (3) are detected by means of pointcut expressions, associated with a device operations' signatures in order to enforce the constraint related to incoming parameter settings and eventually configured to enforce a retry policy.

In all these cases, the fault turns out to be fatal (failures) and requires the component to be reset after saving the data and logging all the additional information, as well as saving the bad parameter setting in order to detect this situation and asking for new parameter values.

In particular, the aspect *FaultDetector* realizes the infrastructure needed to add or remove fault monitoring services on device creation or destruction. Its subaspects provide advice logic for different kinds of fault handling policies, discussed in Section 7.3.1.1. Furthermore, such subaspects define pointcut expressions needed to capture interesting context in which the status of a device must be checked, to perform status decoding, and eventually to send a fault event by means of the logic provided by the classes *FaultListener*.

Figures 8.2 and 8.3 show an excerpt of the code of the *FaultDetector* advice (and its subaspect related to *FastDI* digital integrator device) and of the *DigitalIntegrator_FaultDetector*, respectively.

Twofold analysis were carried out (a) assessing the modularity improvement derived from AOP introduction (internal quality) and (b) evaluating its performance to verify that these benefits do not introduce run-time side effects (external quality).

```
aspect FaultDetector {
public:
        vector<ffmm::core::devices::Virtual_Device*> _mdevs;

        static vector<std::string> showMonitoredDevices()
        static void showMonitoredDevices()
        pointcut devices() = "ffmm::core::devices::Virtual_Device";

        public abstract pointcut retry(Virtual_Device* dev);
        void around(Virtual_Device* dev) : retry(dev);
        advice devices() : slice class {
                virtual init initFaultDetection();
                virtual init removeFaultDetection();
                virtual bool checkDeviceStatus()
                virtual int decodeError()
                virtual void addListener();
                virtual void removeListener();
        }

        void addToMonitor(ffmm::core::devices::Virtual_Device* m);
        void removeFromMonitor(ffmm::core::devices::Virtual_Device* m);
        void printMonitoredDevices();
        void CheckStatus(ffmm::core::devices::Virtual_Device* m)
        advice device_costruction() : after()  {this->CheckStatus(tjp->target());}
        advice device_destruction() : before() {this->removeFromMonitor(tjp->target());}
};
```

Figure 8.2. The abstract *FaultDetector* aspect [6].

```
class FDIListener :
        public IFdiListener<ffmm::core::devices::FastDI>
        {
        public:
                virtual void onFdiError(FdiErrorEvent<ffmm::core::devices::FastDI>* evt);
        };

aspect DigitalIntegrator_FaultDetector  : public FaultDetector {
        public:
                static FDIListener* _lis;
                static FDIListener* getListener() {
                        if (_lis==NULL) _lis = new FDIListener();
                        return FastDI_FaultDetection::aspectOf()->_lis;
                }
        private:
                pointcut devices() = "ffmm::core::devices::FastDI";
                // Concrete slice class for Digital Integrators
                advice devices() : slice class FastDISlice {
                        virtual void initFaultDetection();
                        virtual void removeFaultDetection();
                        virtual bool checkDeviceStatus();
                        virtual int decodeError();
                        virtual void addListener();
                        virtual void removeListener();
                };

        pointcut startOnDigitalIntegrator(FastDI _fdi)=
                call("void FastDI.start(...)") && target(_fdi);
        pointcut stopOnDigitalIntegrator(FastDI _fdi) :
                call("void FastDI.stop(...)") && target(_fdi);

        void around(FastDI _fdi) : startOnDigitalIntegrator(_fdi) {
                if (_fdi._status.started) callNotifier(…)
                else proceed(_fdi);
                if (check_postconditions())
                        // multicast fault to the listener via fire% methods
                        callNotifier(new FastDIFaultDecoder(_fdi).getTable());
        }

        void around(FastDI _fdi) : stopOnDigitalIntegrator(_fdi) {
                if (_fdi._status.stopped) callNotifier(…)
                else proceed(_fdi);
                if (check_postconditions())
                        // multicast fault to the listener via fire% methods
                        callNotifier(new FastDIFaultDecoder(_fdi).getTable());
        }
        …
};
```

Figure 8.3. Excerpt of *DigitalIntegrator_FaultDetector* [6].

8.2.4 MODULARITY COMPARISON

The software quality of the AOP version of the *Fault Detector* was evaluated by comparing it with the corresponding OOP version that previously existed inside the FFMM. With this aim, the software quality attribute of modularity was assessed for both the AOP and OOP versions by evaluating (a) the percentage of lines of source code related to fault detection logic present in each module with respect to the total lines of code (LOC) of the same module and (b) the Degree of Scattering (DOS) and Degree of Focus (DOF) metrics [7], for each module and fault detection concern. The analysis was focused on the most relevant fault sources of the FFMM, that is, the modules implementing devices.

In Table 8.1, the analysis results along with the ratio of code duplicated in the different software modules (cloned code ratio) are reported. In the OOP version of the *Fault Detector*, a high level of cloned code exists, because in each device's operation often the same tests against the internal status are requested. Conversely, in the AOP version, this ratio is drastically reduced. An ideal implementation of the fault detection concern would have a null DOS and a DOF equal to one, for each device module (i.e., each device is focused only on the base concern and does not contribute at all to the *Fault Detector* concern). Table 8.1 shows that the OOP version has a very-low value of DOF, for each module (i.e., all modules contribute to the fault detection concern), and a DOS for the fault detection concern near to the maximum (uniformly scattered). This means that the fault detection concern in the OOP version has absolutely unacceptable values of modularity.

Thus, a not trivial maintenance (or evolution) is very difficult, because each modification could affect and require changes in many different software modules (i.e., mainly all device modules).

Table 8.1. Fault detection code in each device module and computation of percentage DOF and DOS metric for both OOP and AOP versions (OOP: Object-Oriented Programming; AOP: Aspect-Oriented Programming; LOC: Lines of code; DOF: Degree of focus; DOS: Degree of scattering)

Device	OOP FD % LOC	OOP Cloned % LOC	OOP DOF	OOP DOS	AOP % LOC	AOP DOF	AOP DOS
FastDI	15.75	10.93	0.17		0.81	0.97	
Maxon Epos	18.04	9.78	0.21		0.73	0.97	
EncoderBoard	21.53	14.27	0.28	0.96	0.97	0.98	0.13
PowerSupply	18.36	12.70	0.24		0.97	0.90	
Transducer	21.15	8.24	0.27		0.97	0.93	
Keithley2k	18.48	11.32	0.16		0.97	0.97	

Instead, in Table 8.1, DOS values of the AOP version are near to the minimum: The fault detection concern is well modularized in one module (the *FaultDetector* aspect), and each device module is marginally involved in the concern (such as mentioned before, this is due to the fault and error broadcasting methods not yet removed from the devices). This result is better highlighted in Figures 8.4 and 8.5. In particular, in Figure 8.4, the percentage LOC (%LOC) of the Fault Detection concern for all the device modules are compared for both AOP and OOP versions. In the figure, the ratio of the cloned LOCs in the OOP implementation, completely removed

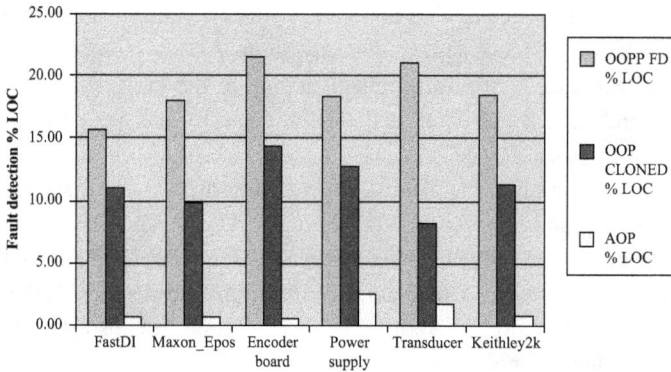

Figure 8.4. Percentage lines of code (LOC%) of fault detection concern in device modules for OOP and AOP versions.

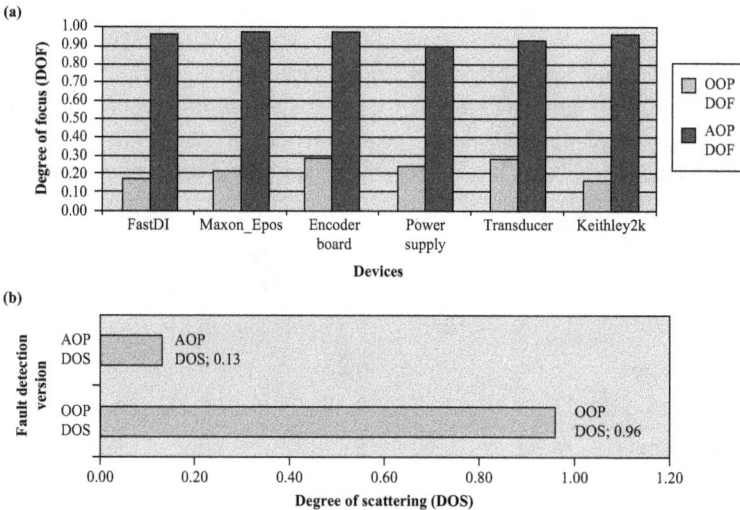

Figure 8.5. (a) DOS and (b) DOF comparisons of OOP and AOP versions with respect to fault detection concern.

in the AOP version, is reported. Of course, the cloned code makes worse the maintainability and increases the probability of introducing bugs in the code.

In Figure 8.5, the level of DOF (a) and DOS (b) for each device module with respect to Base System and Fault Detection concerns is reported. The results show a radically increased modularity for the AOP version, because each device module is much more focused on the base concern with respect to the OOP version. Moreover, the fault detection concern is highly scattered in the OOP version (high values of DOS), while it is very focused in the AOP implementation (very low values for DOS).

8.2.5 PERFORMANCE VERIFICATION

The analysis was aimed at verifying experimentally that the AOP architecture would not have a negative impact on run-time performance of the overall system (due to aspect runtime interception overhead). With this intent, the AOP system was instrumented to gather execution times of the aspect overheads. In both versions, fault detection times related to fault decoding and handling, are present. They were filtered out from the analysis. Therefore, main attention was paid to evaluate the overheads added by the AOP interception mechanism to the fault detection time in order to assess the effectiveness of the AOP architecture, that is, that the AOP response times are not worse than the OOP version. The afore-described analysis was carried out by running two versions of the software in the same conditions. The runs were performed by causing some previously described faults in the measurement station. Those faults were induced intentionally in different ways, for example:

- by providing devices with wrong parameter values,
- by interrupting the communication between the PC and the devices (device not found, or communication timeout if the communication with device had already been established),
- by starting the FDI acquisition procedure without feeding the instrument with the required trigger signal (measurement timeout),
- and by adding a delay in the execution of some commands (command timeout).

The worse average times in several different categories of fault detection pointcut expressions (i.e., device creation/destruction, interception of device operations) were selected, and the time spent in the aspect runtime to jump to fault detection routines was collected. These are reported in the last column of Table 8.2 (in percentage of the total time spent in the aspect). For the sake of clarity, the results of Table 8.2[1] are reported in Figure 8.6.

Times needed to handle creation and destruction of devices (pointcut expressions from rows 1 to 8) are greater than those required to handle faults during measurement tasks (pointcut expressions from rows 9 to 15). In the former cases, the fault detector infrastructure must be set up for devices being created. This requires more time than the other kind of pointcut expressions, which have only to capture the context of an operation, issuing a fault event if necessary. These times are comparable to those of the OOP version, where listeners are explicitly registered with the created devices to handle faults. In these cases, the aspect overhead is particularly reduced with respect to the entire fault detection tasks. Pointcut expressions ranging from row 9 to row 15 are related to fault detection during normal device operations. Their goal is to capture the entire context in which the device state changes to check its validity. In the developed AOP implementation, the worst overheads due to aspect interception mechanism (see the last column in Table 8.2) are always less than 1.5% of the fault detection times (the worst case is for the interception of calls to *EncoderBoard* device operations with complex arguments matching expression to check preconditions; related to pointcut expression at row 9). Therefore, the suitability of such performance overhead in the concrete measurement scenario was assessed, and all the timing constraints were satisfied.

8.2.6 DISCUSSION

The AOP architecture allows a high level of flexibility by performing very complex and bendable run-time binding among sources and handlers of the faults, without affecting significantly the performance, while keeping the detection code well modularized in its hierarchy. Another main advantage of such a technique is the maintainability and the reusability of the code: for each new device added to the framework, the related fault detection code is added to the fault detection hierarchy. Since all fault detection code is well modulated in few subaspects, commonalities among different fault detection logic are well structured and factored out. As a consequence, the *FaultDetector* design, with respect to "traditional" OOP

---

[1]The times refer to a Pentium IV 1.3 GHz machine, with 512 Mb of RAM running the instrumented AOP version.

Table 8.2. Worst average times spent in aspect runtime with respect to device creation and destruction and fault detection point cuts

	Pointcut expressions	Total Time (ms) spent in aspect	Time (ms) spent in matching advices	Time (ms) spent in aspect runtime	%Time spent in aspect runtime
1	EncoderBoard construction	13,029641	13,028840	0,000801	0,006%
2	FastDI construction	45,912234	45,893478	0,018756	0,041%
3	FDICluster (1 element) construction	59,062982	59,010519	0,052463	0,089%
4	Maxon–Epos construction	14,013891	14,011993	0,001898	0,014%
5	EncoderBoard destruction	10,282883	10,282652	0,000231	0,002%
6	FastDI destruction	18,511722	18,511302	0,000420	0,002%
7	FDICluster (1 element) destruction	6,034312	6,034096	0,000216	0,004%
8	Maxon–Epos destruction	11,045256	11,045013	0,000243	0,002%
9	within(EncoderBoard)&&call(%) && args(...)	2,8803261	2,842784	0,037542	1,303%
10	withincode(FastDI) && call(plx->write)	1,146675	1,144412	0,002263	0,197%
11	withincode(FastDI) && call(plx->read)	1,486675	1,476675	0,010000	0,673%
12	withincode(Maxon–Epos) && execution(set%)	0,982102	0,980102	0,002000	0,204%
13	withincode(Maxon–Epos) && execution(get%)	0,902345	0,899015	0,003330	0,369%
14	withincode(Maxon–Epos) && execution(start())	2,896735	2,886735	0,010000	0,345%
15	withincode(Maxon–Epos) && execution(stop())	2,416375	2,406875	0,010000	0,414%

(a)

(b)

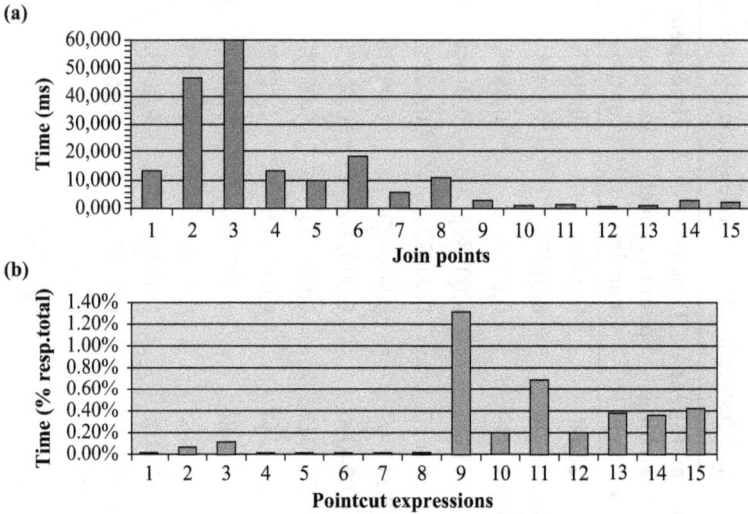

Figure 8.6. (a) Total average and (b) worst case overhead times spent in aspect runtime. The pointcut expressions numbering refers to Table 8.2.

version, exhibits a much more centralized design, greatly increasing the possibility of code reuse, and reducing code duplication [6]. Finally, the AOP architecture is not targeted at a specific system component, and the same fault detector architecture can be reused to detect different kinds of faults in different components.

8.3   SYNCHRONIZER [8]

The Petri-net *Synchronizer* is validated in a magnetic permeability measurement system by highlighting the twofold expected advantages of the simplification of the measurement script and the speeding up of its definition. In the following, (a) the *case study on magnetic permeability* and (b) the corresponding *measurement procedure*, exploiting the Petri net-based *Synchronizer*, are illustrated.

8.3.1   CASE STUDY ON MAGNETIC PERMEABILITY MEASUREMENT

In the following, a case study on the method of the split-coil permeameter [9] for measuring the magnetic permeability is described. The split-coil permeameter is composed of two coils wound in a toroidal shape, which can be opened allowing a toroidal specimen of the material under test to

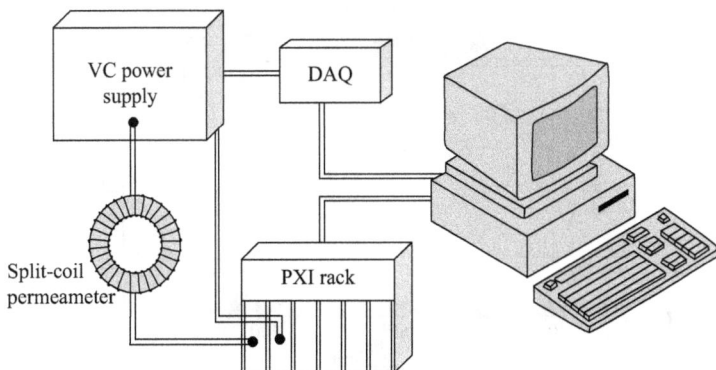

Figure 8.7. Layout of the split-coil permeability measurement setup [8].

be wrapped. One coil is needed to excite the field and the other one to capture the flux.

A PC (Figure 8.7), hosting the FFMM with the *Synchronizer*, is linked to a data acquisition board (DAQ [10]), in order to control the Voltage Controlled (VC) Power Supply to the excitation coil of the split-coil permeameter by the analog output. A PXI crate containing

- a Fast Digital Integrator (FDI [4]), a CERN proprietary general-purpose digital board, configured for the coil signal acquisition and numerical integration;
- a CERN proprietary trigger board, for managing and feeding the trigger input of the FDIs;
- a further FDI, to acquire the excitation current and the relative flux linked to the measurement coil, and to generate the trigger for the acquisition by the trigger board.

8.3.2 MEASUREMENT PROCEDURE

The specimen is magnetized gradually by using a current waveform (Figure 8.8) made by a series of linear ramps and plateau with exponentially increasing amplitude (cycles). A current cycle is composed of an initial plateau, a linear ramp with constant ramp rate, and a final plateau [8].

After the devices' setup, the measurement algorithm is composed of the following steps [8]:

1. Demagnetization of the specimen [9]
2. Acquisition of current and flux

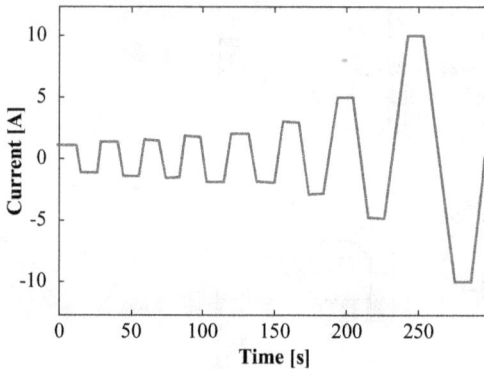

Figure 8.8. Current cycles [8].

```
ADD_TASK(Demagnetization)
ADD_TASK_AFTER_TASK(Demagnetization, Set_Next_Cycle)
ADD_TASK_AFTER_EVENT(next_cycle, Current_Cycle)
ADD_TASK_AFTER_EVENT(start_cycle, Start_Acquisition)
ADD_TASK_AFTER_EVENT(stop_cycle, Stop_Acquisition)
ADD_TASK_AFTER_TASK(Stop_Acquisition, Set_Next_Cycle)
ADD_TASK_AFTER_ EVENT(end_measurement, Data_Conversion)
```

Figure 8.9. FFMM script fragment defining the *Execution Graph* [8].

3. Generation of one cycle of the signal controlling the power supply
4. Waiting for the completion of the actual current cycle
5. Stopping the acquisition of the flux
6. Starting the generation of the next current cycle and going to 3 or, if the maximum value of current is reached, stopping the acquisition
7. Data conversion

In such a measurement procedure, several tasks have to be executed with time constraints. This is introduced easily into a FFMM script by using the new features of the *Synchronizer*. The test engineer first has to write the short high-level procedures describing each step of the whole measurement algorithm in separate tasks by exploiting the related tools of the FFMM [11]. Then, for the synchronization, he adds each task to the execution tree effortlessly, such as shown in Figure 8.9, without worrying about the time synchronization of parallel or series tasks.

As an example, to add a task to be executed after another one, the test engineer has to use the statement *ADD_TASK_AFTER_TASK(first_task, second_task)* such as shown in the case study (Figure 8.9) to the *Set_Next_Cycle* task, scheduled in series to the *Demagnetization* task. Otherwise, if the test engineer wants to arrange the start of the execution of one task after another task's event and, then, to let both tasks be active in parallel, he has to exploit the statement *ADD_TASK_AFTER_EVENT(event, second_task)* like shown in Figure 8.9, where the *Current_Cycle* task is scheduled after the event *next_cycle*.

The measurement algorithm, previously described step by step, is codified in the script, and the related *Execution Graph* is shown in Figure 8.10 [8].

The first task to be active is the *Demagnetization*, at the end of this task the execution passes to the *Set_Next_Cycle* task (series execution); *Set_Next_Cycle* sets up the actual cycle of current and starts the *Current_Cycle* by throwing the event *next_cycle; Current_Cycle* first starts the task *Start_Acquisition* that enables the acquisition of flux and current, then begins the generation of the actual current cycle, then stops the acquisition by throwing the event *stop_cycle*; the task *Stop_Acquisition* triggers a new execution of *Set_Next_Cycle* that restarts the loop up till the last scheduled current cycle or enables the *Data_Conversion* in order to format the output data.

In Figure 8.11, the result of the measurement is shown graphically, by referring to a series of current cycles starting from 0 up to 10 *A* and their corresponding magnetic field from 0 to 6000 *A/m*. The right trend of the hysteresis curve of the iron specimen highlights the validity of the proposal.

8.3.3 DISCUSSION

Petri nets have shown to be an appropriate formalism in managing asynchronous measurement tasks scheduling. The method presented in this book turns out to be well suited in assuring a proper software synchronization of the test procedure because it allows the temporal dimension to be abstracted from a flat description of the sequences of events, by discovering the actual temporal relations between the events. In particular, the *Synchronizer* allows a test engineer, without software skill, to schedule concurrent, sequential, and event-based tasks, with an intuitive approach, by thinking, into the time domain, in terms of a simple relation like "after that" or "at the same time of."

This approach was tested on the field in the permeability measurement of the FFMM at CERN. The main advantage is the simplification of the measurement script and the speeding up of its design. As a consequence, the test engineer can concentrate his attention on the measurement

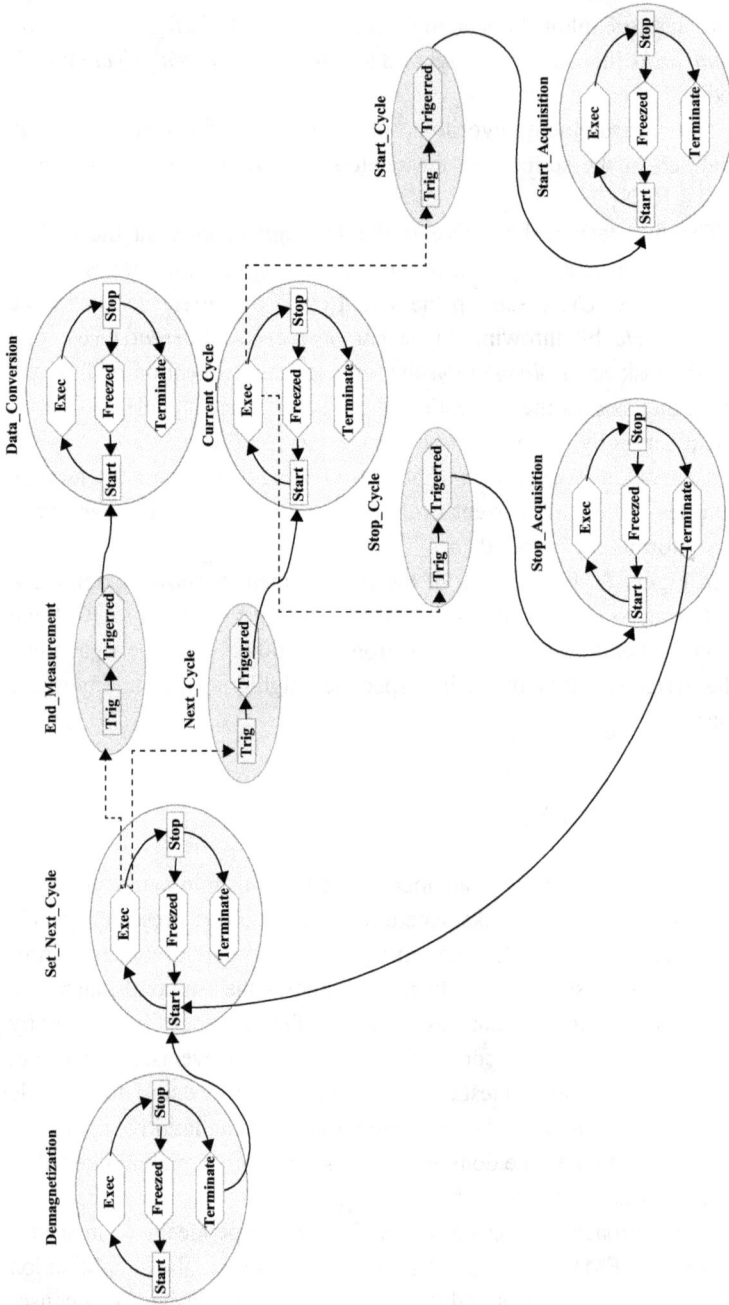

Figure 8.10. Execution graph of the case study on permeability measurement [8].

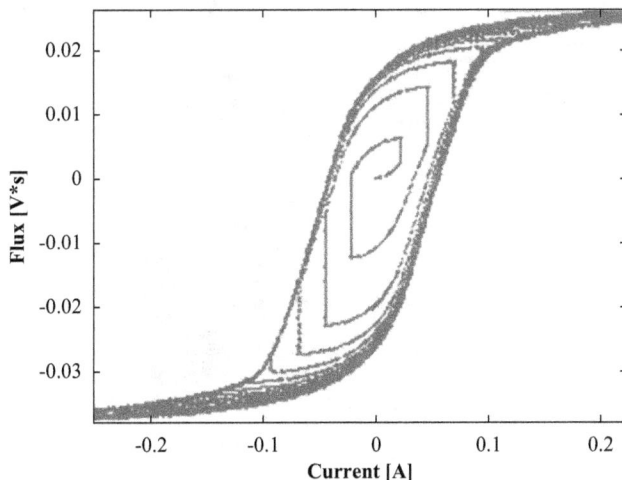

Figure 8.11. Hysteresis curve of the material [8].

matters and not worry about software design and programming details like threads and semaphores. The corresponding, satisfying results motivated a wide use of the *Synchronizer* in other measurement layouts at CERN.

8.4 DOMAIN SPECIFIC LANGUAGE

An experimental case study, conceived specifically for testing the Measurement Domain Specific Language (MDSL) in an actual measurement application, was envisaged in the frame of the FFMM [5] at CERN. The main objective is to verify on the field its benefits in terms of simplicity, effectiveness, and flexibility in the measurement software applications production.

In the following, the MDSL implementation and application results are highlighted by referring to two practical examples: superconducting magnet testing and permeability measurements.

8.4.1 CASE STUDY ON SUPERCONDUCTING MAGNET TESTING [12]

In the example, the data are treated as 32-bit floating point and stored in ASCII format in order to simplify the exchange among different tools.

In Figure 8.12, the MDSL script as a whole for the procedure of superconducting magnet test already described in Section 8.2 is shown. The

```
BEGIN_SCRIPT "Superconducting Magnet Bench":

    BEGIN_TASK "Devices_definition":
        DEF FDI_CLUSTER: "FDI_Cl_1" WITH ("FDI_Cl_1");
        DEF ENCODER_BOARD: "Enc_Brd " WITH ( "Enc_Brd" );
        DEF MAXON_EPOS: "mtrl" WITH ( "Mtrl" );
        DEF POWER_CONTROLLER: "Pow_Ctr" WITH ( "Pow_Ctr" );
        DEF TIMING_BOARD: "Time_Brd " WITH ( "Time_Brd" );
    END_TASK

    BEGIN_TASK "Devices_configuration":
        CFG FDI_CLUSTER: "FDI_Cl_1" WITH ([12,12,12,13,13,13],[8,9,10,8,9,10]);
        CFG ENCODER_BOARD: "Enc_Brd" WITH ( 11 , 8 );
        CFG MAXON_EPOS: "Mtrl" WITH ( 6, 100 );
        CFG POWER_CONTROLLER: "Pow_Ctr" WITH ( "cfc-sm18-r2",5001,50);
        CFG TIMING_BOARD: "Time_Brd" WITH ( 11,9,1000 );
    END_TASK

    BEGIN_TASK "Devices_setting":
        SET FDI_CLUSTER: SetParams   ("FDI_Cl_1",512,0.1,20,CONT,500000,2048,100);
        SET ENCODER_BOARD: SetParams ("Enc_Brd", 1, 512, 1, 0);
        SET MAXON_EPOS: SetParams("Mtrl", 0, 10 , 10 , 60);
        SET TIMING_BOARD: TimeStampTrigger_PF1("Time_Brd");
    END_TASK

    BEGIN_TASK "Measurement":
        CMD MAXON_EPOS: Start ("Mtrl");
        CMD MAXON_EPOS: Wait ("Mtrl");
        CMD POWER_CONTROLLER: Startlog("Pow_Ctr", "c:\\current_data.txt");
        CMD FDI_CLUSTER: StartAcquisition ("FDI_Cl_1", "c:\\dfluxes1_data.txt");
        CMD ENCODER_BOARD:StartTrigger ("Enc_Brd", 1 );
        CMD TIMING_BOARD: ReadTimeStamp("Time_Brd",1);
        CMD TIMING_BOARD: DisableTimeStamp("Time_Brd" );
        WAIT 1000 ms;
        CMD FDI_CLUSTER: StopAcquisition ("FDI_Cl_1");
        CMD FDI_CLUSTER: WaitAcquisition ("FDI_Cl_1");
        CMD ENCODER_BOARD: Stop_Trigger ("Enc_Brd", 1 );
        CMD MAXON_EPOS: Stop("Mtrl");
        CMD MAXON_EPOS: WaitStop("Mtrl");
        CMD POWER_CONTROLLER: Stoplog("Pow_Ctr");
    END_TASK

    ADD_TASK (Devices_definition)
    ADD_TASK_AFTER_TASK (Devices_configuration,Devices_definition)

    ADD_TASK_AFTER_TASK (Devices_definition, Devices_setting)

    ADD_TASK_AFTER_TASK (Devices_setting, Measurement)

END_SCRIPT
```

Figure 8.12. Superconducting magnet test script [12].

script is composed by four tasks: (a) "*Devices_definition*," for pointing out the devices to be used [3]; (b) "*Devices_configuration*," where the device connections are configured; (c) "*Devices_setting*" for selecting the measurement parameters; and (d) "*Measurement*," where the actual measurement is started and stopped. In particular, the measurement is carried out by using five different devices: (a) a set of FDIs (FDI_CLUSTER), (b) an encoder board (ENCODER_BOARD), (c) a motor controller (MAXON_EPOS), (d) a power generator (POWER_CONTROLLER), and (e) timing board (TIMING_BOARD). In the device definition task, each instrument is declared with a unique name as a string. In the device configuration and setting tasks, the configuration and setting methods of the already defined devices are called with the requested parameters suggested by the IDE during the script writing (usually strings or numbers). The measurement task contains the commands related to the actions of the devices (Start, Wait, Stop, and so on) and usually their argument is the device unique name as string. After the definition, each task is added to the execution tree [8], where in this case only a sequential run is needed.

Each task section contains a small number of methods (less than 15 usually). In general, these methods are implemented into the C++ class controlling the specific device. MDSL inherits this kind of methods from host framework classes, and the benefit of the new language is related to the suggestion and the check of the arguments to be passed and the right use of the device methods in the script, for example, for *MAXON_EPOS* motor device: *Wait* can be used only after *Start*, and *Start* can be used only after *Set*.

In Figure 8.13, the measured integral sextuple component is shown as a function of the time and current. In both plots, the data are scaled by a constant value in order to make comparable the curves and to emphasize the "decay" and "snapback" phenomena [13], typical in superconductive magnets. In particular, the measurements at 2 kA and 6 kA, having common powering history parameters, are shown. The plots are obtained after a data postprocessing by using other tools (MATLAB®).

The MDSL script for superconducting magnet test emphasizes the power of the method presented in this book. With less than 50 code lines (a compression ratio of 1:10 compared to C++ code) organized in few coherent sections, a whole measurement application involving several different devices can be produced. By taking into account (a) the decrease of the number of code lines and (b) the help provided by specific grammar rules embedded into the IDE, measurement software applications turn out to be easier to produce than in standard programming languages.

8.4.2 CASE STUDY ON MAGNETIC PERMEABILITY MEASUREMENT

In the following, the MDSL script performing the permeability measurement by means of the split-coil permeameter described in Section 8.3 [14] is presented.

The measurement algorithm described earlier in Section 8.3.2 is codified in the script in six tasks. In Figure 8.14, the script structure is shown. The commands in each task are omitted in order to emphasize the script structure and make it more readable.

The first task to be active is the *Demagnetization*, at the end of this task the execution passes to the *Set_Next_Cycle* task (series execution); *Set_Next_Cycle* sets up the actual cycle of current and starts the *Current_Cycle* by throwing the event *next_cycle; Current_Cycle* first starts the task *Start_Acquisition* that enables the acquisition of flux and current, then begins the generation of the actual current cycle, then stops the acquisition by throwing the event *stop_cycle*; the task *Stop_Acquisition* triggers a new

(a)

(b)

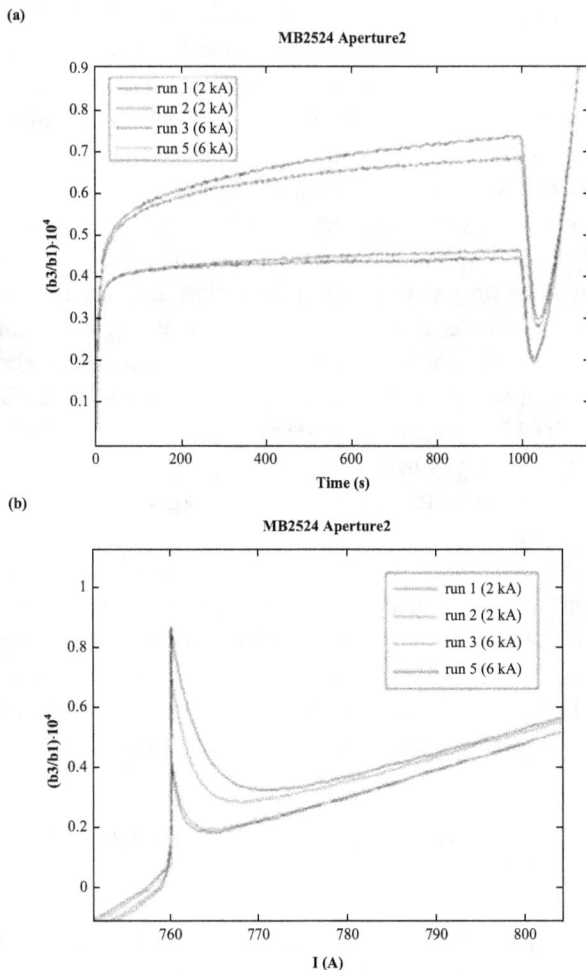

Figure 8.13. Measured normal sextupolar "*decay*" and "*snapback*" as a function of (a) the time and (b) as a function of the measured current for different supply current cycles (data are scaled to be compared).

execution of *Set_Next_Cycle* that restarts the loop up till the last scheduled current cycle or enables the *End_Test* to set in stand by the devices.

The difference from the previous example is that in this case the task execution tree is more complicated, and it is composed of serial and parallel running tasks started by events. Simultaneous tasks can be managed in the script at a very high level of abstraction. Race conditions and execution tree

```
BEGIN_SCRIPT "Permeability Bench":

    BEGIN_TASK "Devices_definition":
                    ...
    END_TASK

    BEGIN_TASK "Demagnetization":
                    ...
    END_TASK

    BEGIN_TASK " Set_Next_Cycle ":
                    ...
    END_TASK

    BEGIN_TASK " Start_Acquisition ":
                    ...
    END_TASK

    BEGIN_TASK " Stop_Acquisition ":
                    ...
    END_TASK

    BEGIN_TASK " End_Test ":
                    ...
    END_TASK

    ADD_TASK (Devices_definition)
    ADD_TASK_AFTER_TASK (Demagnetization, Devices_definition)
    ADD_TASK_AFTER_TASK (Set_Next_Cycle, Demagnetization)
    ADD_TASK_AFTER_EVENT   (Current_Cycle, next_cycle)

    ADD_TASK_AFTER_EVENT   (Start_Acquisition, start_cycle)

    ADD_TASK_AFTER_EVENT   (Stop_Acquisition, stop_cycle)

    ADD_TASK_AFTER_TASK (Set_Next_Cycle, Stop_Acquisition)

    ADD_TASK_AFTER_EVENT   (End_test, end_measurement)

END_SCRIPT
```

Figure 8.14. Permeability measurement MDSL script [12].

of serial and parallel tasks in the script have only to be described by using the task-related commands (ADD_TASK, ADD_TASK_AFTER_TASK, ADD_TASK_AFTER_EVENT). The control of the race conditions is delegated to a specific software component, that is, the software synchronizer, implemented in the host language [8]. The method presented here allows the production of complex measurement applications by a composition of simple steps: test engineer after the definition of atomic tasks can arrange them into an elaborate layout without increasing the script complexity.

In Figure 8.15, the results of two measurements of the same steel specimen are shown: The effects of eddy currents can be appreciated at different current ramp rates. The plots are obtained after a data postprocessing by using other tools (MATLAB).

(a)

(b)

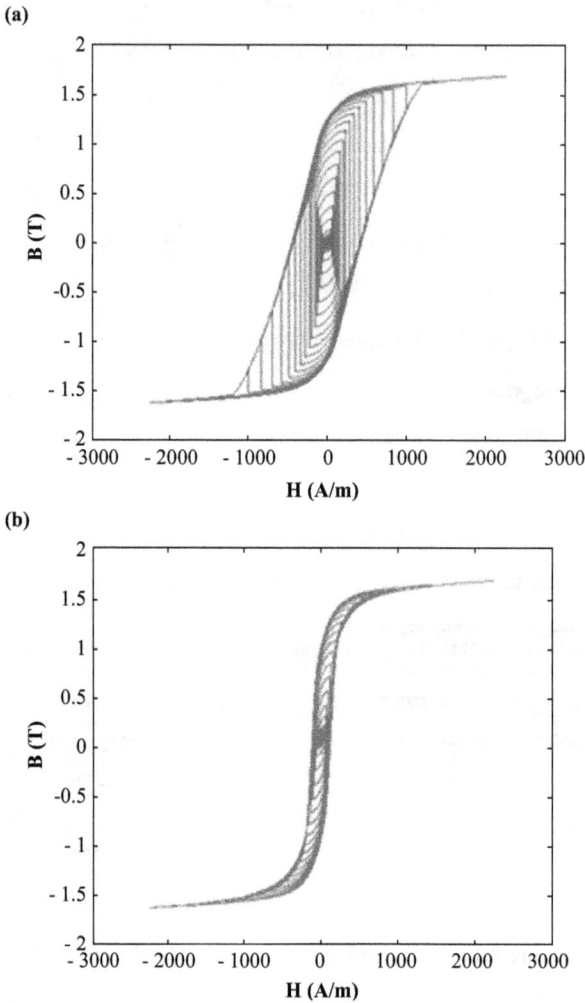

Figure 8.15. Permeability measurement results for different current ramp rates: (a) 0.5 A/s and (b) 0.01 A/s.

8.4.3 DISCUSSION

A DSL with specialized constructs concerning the automation of measurement procedures is presented. The technique is based on the Model-Driven Engineering concepts and DSL solutions. Basic idea is to provide unskilled programmers with means of producing concise and bug-free specific measurement applications by using scripts written in a high-level and domain-representative language, with the effect of improving the resulting applications and reducing the development effort. Moreover, a set of domain rules can be embedded in the language definition with a straightforward advantage in preventing errors during the early phases of the development. The on-field applications of the new technique point out the benefits in terms of simplicity, effectiveness, and flexibility in the measurement software applications production, and the positive feedbacks from users are fostering a wider use.

8.5 ADVANCED USER INTERFACES GENERATOR [15]

This section is focused on highlighting how the approach presented in this book supports test engineers in generating a GUI, automatically, during a usual FFMM operation at CERN [6, 16]. FFMM, in addition to satisfying all the functional [15] requirements, can provide means to generate the graphical user interface.

Next, (a) the *case study of magnetic permeability measurement* and (b) the *measurement results* are reported.

8.5.1 CASE STUDY ON MAGNETIC PERMEABILITY MEASUREMENT

The case study is tasked at highlighting how the approach presented in this book supports test engineers in generating the GUI automatically for a measurement procedure based on the methods of the split-coil permeameter illustrated in Section 8.3.1 (Figure 8.7) [14]. The procedure is codified in the application script and processed by the FFMM framework in order to produce an executable file. The devices involved in the measurement procedure are configured by means of the automatically-generated user interface (AUI) features of FFMM [15].

As an example, at the beginning of the measurement script, the test engineer needs to configure the FDIs: the number of FDI and their bus are required to start the acquisition. Thus, the test engineer enters the following statements in to the script [15]:

- Def GIC "*InputParam*";
- Capture *InputParam* with (*numFDI* 1, "Parameter Request:", "number of FDI");
- Capture *InputParam* with (*bus*, *numFDI*, "Parameter Request:", "FDI bus").

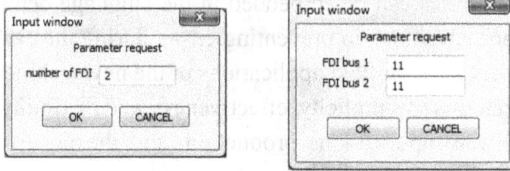

Figure 8.16. FDI configuring forms [15].

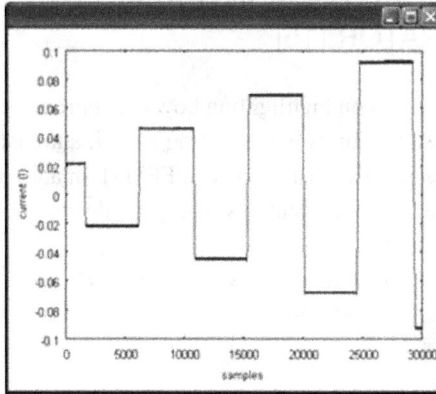

Figure 8.17. A window plotting some current cycles [15].

Then, during the application execution, the forms are generated (Figure 8.16).

As a further example, during the measurement, the test engineer can program the application to show the user the current flow by using the plot feature of the GIC object. Thus the following statements have to be placed in the script:

- Def GIC *"CurrentPlot"*
- Plot *CurrentPlot* with (*currentData*)

During the application execution, the window with the plot is generated as shown in Figure 8.17 [15].

A steel specimen was tested and, according to the afore-explained procedure [14], the permeability characteristic curve may be obtained by analyzing the data.

By validating the AUI using the Model-View-Interactor paradigm, the relative magnetic permeability versus the magnetic field curve is reported in Figure 8.18.

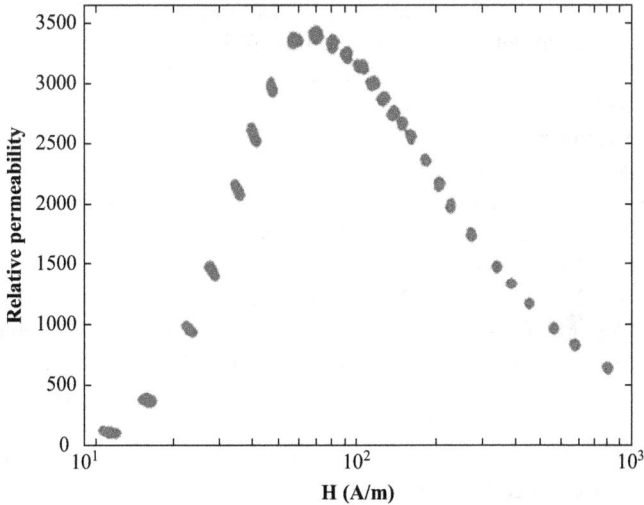

Figure 8.18. Relative permeability versus magnetic field curve [15].

8.5.2 DISCUSSION

The Model-View-Interactor paradigm for AUI is mainly for test engineers using FFMM to easily produce the GUI for their measurement applications. The advantages of the technique presented in this book meet the requirements of software framework for measurements systems and agree with the basic idea, primarily by decreasing the performance and cost ratio of the application development even with a graphical interface.

REFERENCES

[1] Bottura, L. April 11–17, 1997. "Field Dynamics in Superconducting Magnets for Particle Accelerators." *CERN Accelerator School on Measurement and Alignment of Accelerator and Detector Magnets*, pp. 79–105. Anacapri, Italy: CERN

[2] Arpaia, P., L. Bottura, L. Fiscarelli, and L. Walckiers. 2012. "Performance of a Fast Digital Integrator in on-Field Magnetic Measurements for Particle Accelerators." *Review of Scientific. Instruments* 83, no. 2, article id. 024702. doi: http://dx.doi.org/10.1063/1.3673000

[3] Bottura, L., and K.N. Henrichsen. 2004–2008. "Field measurements." *CERN Accelerator School*, p. 118. Erice, Italy: CERN report.

[4] Arpaia, P., A. Masi, and G. Spiezia. April 2007. "A Digital Integrator for Fast and Accurate Measurement of Magnetic Flux by Rotating Coils." *IEEE*

Transactions on Instrumentation and Measurement 56, no. 2, pp. 216–20. doi: http://dx.doi.org/10.1109/tim.2007.890787.

[5] Arpaia, P., M. Buzio, L. Fiscarelli, and V. Inglese. November 2012. "A Software Framework for Developing Measurement Applications Under Variable Requirements." *AIP Review of Scientific Instruments* 83, no. 11, article id. 115103. doi: 10.1063/1.4764664

[6] Arpaia, P., M.L. Bernardi, G.Di Lucca, V. Inglese, and G. Spiezia. 2010. "An Aspect-Oriented Programming-Based Approach to Software Development for Fault Detection in Measurement Systems." *Computer Standards & Interfaces* 32, no. 3, pp. 141–52. doi: http://dx.doi.org/10.1016/j.csi.2009.11.009

[7] Eaddy, M., T. Zimmermann, K.D. Sherwood, V. Garg, G.C. Murphy, N. Nagappan, and A.V. Aho. 2008. "Do Crosscutting Concerns Cause Defects?" *IEEE Transactions on Software Engineering* 34, no. 4, pp. 497–515. doi: http://dx.doi.org/10.1109/tse.2008.36

[8] Arpaia, P., L. Fiscarelli, G. La Commara, and F. Romano. January 2011. "A Petri Net-Based Software Synchronizer For Automatic Measurement Systems." *IEEE Transactions on Instrumentation and Measurement* 60, no. 1, pp. 319–28, doi: 10.1109/TIM.2010.2046602

[9] Henrichsen, K.N. 1967. "Permeameter." In *Proceeding second Internationa. Conference on Magnet Technology*, pp. 735–9. Oxford, U.K: Rutherford Laboratory.

[10] National Instruments. 2010. DAQ, http://www.ni.com/data-acquisition/

[11] Arpaia, P., M. Buzio, L. Fiscarelli, V. Inglese, and G. La Commara. May 5–7, 2009. "Measurement-Domain Specific Language for Magnetic Test Specifications at CERN." In *Proceeding. IEEE I2MTC*, pp. 1716–20. Singapore: IEEE.

[12] Arpaia, P., L. Fiscarelli, G. La Commara, and C. Petrone. 2011. "A Model-Driven Domain-Specific Scripting Language for Measurement System Frameworks." *IEEE Transactions on Instrumentation and Measurement* 60, no. 12, pp. 3756–66. doi: 10.1109/TIM.2011.2149310

[13] Bottura, L., L. Walckiers, and R. Wolf. June 1997. "Field Errors Decay and 'Snapback' in Lhc Model Dipoles." *IEEE Transactions Applied Superconductivity* 7, no. 2, pp. 602–5. doi: http://dx.doi.org/10.1109/77.614576

[14] Arpaia, P., M. Buzio, L. Fiscarelli, G. Montenero, and L.Walckiers. May 3–6, 2010. "High Performance Permeability Measurement: A Case Study at CERN." In *Proceeding of the International Instrumentation and Measurement Technology. Conference*, pp. 58–61. Austin, TX: IEEE.

[15] Arpaia, P., L. Fiscarelli, and G. La Commara. 2010. "Advanced User Interface Generation in the Software Framework for Magnetic Measurements at CERN." *Metrology and Measurement Systems* 17, no. 1, pp. 27–38. doi: http://dx.doi.org/10.2478/v10178-010-0003-y

[16] Arpaia, P., M. Bernardi, G. Di Lucca, V. Inglese, and G. Spiezia. May 12–15, 2008. "Aspect Oriented-based Software Synchronization in Automatic Measurement Systems." *Instrumentation and Measurement Technology Conference Proceedings*, pp. 1718–21. Victoria, BC: IEEE.

CHAPTER 9

FRAMEWORK VALIDATION ON LHC-RELATED APPLICATIONS

It's supposed to be automatic, but actually you have to push this button.
—John Brunner, Stand on Zanzibar

9.1 OVERVIEW

At CERN, in the last few years, the Flexible Framework for Magnetic Measurements (FFMM) [1] has been used to develop applications for several test activities. These scenarios have been a valid test bed for checking the FFMM capability of offering an environment for a fast development of different measurement applications with disparate requirements. A specific discussion of the goal achievement in functional terms was presented in Chapter 8 for main FFMM components. In this chapter, for some case studies typical of the CERN magnetic measurements, the attention is focused on the application software as a whole. For each case study, the measurement procedure, the test station, and the experimental results are illustrated. In particular, the applications for measuring the magnetic permeability by means of the split-coil permeameter and the magnetic field by means of the rotating coils, as well as for testing and compensating the field distortion of the superconducting cryo-magnets of the Large Hadron Collider (LHC), are illustrated. Finally, a specific assessment of the flexibility of the FFMM is presented on the basis of the method presented in Chapter 5. The impact of adding and modifying a device, changing service strategies, and implementing new measurement algorithms is assessed.

9.2   ON-FIELD FUNCTIONAL TESTS

In the following, three case studies of FFMM exploitation at CERN are presented:

1. The first case aims at measuring the magnetic permeability of a material sample through a fixed coil transducer.
2. The second application is based on the rotating coil technique and is focused on assessing the quality field of the superconducting cryo-magnets of the LHC.
3. In the third case study, the field errors due to nonideality of the LHC superconducting dipoles are estimated and compensated for the *tracking tests*, that is, the experiments for validating the synthetic magnetic model of the machine.

9.2.1   MAGNETIC PERMEABILITY MEASUREMENTS

Magnetic permeability measurements are of main interest in order to exploit the properties of materials to improve accelerator technologies. In particular, for the LHC, it is important to characterize the magnetic properties of the laminated low-carbon steel used for the magnet yokes.

In practice, toroidal specimens are used for magnetic permeability measurements of soft materials in order to avoid test problems related to bars or strips samples, namely end-effects and gaps or joints in the magnetic circuit [2]. Furthermore, a ring-shaped specimen allows the mean magnetizing force to be computed accurately from a measurement of the magnetizing current, the dimensions of the specimen, and the number of turns in the magnetizing windings. Therefore, this kind of a specimen is the closest to the ideal case when considering the testing principle, even if it's not suitable when only just end usage aspects have to be investigated.

The split-coil permeameter built and used for the magnetic property characterization at CERN is shown in Figure 9.1.

The permeameter consists of three toroidal windings, which can be opened for placing the sample. The two outer coils form the 180-turn excitation winding and the inner 90-turn coil the flux measurement winding. The maximum excitation current, passing through the coil, is limited to 40 A in order to avoid overheating, thus the maximum magnetizing field is approximately 24,000 A/m at room temperature.

Acquisition systems developed in the past before the FFMM provided a low sampling rate of the hysteresis curve for the new generations of

Figure 9.1. Split-coil permeameter [3].

magnetic materials under test, as well as a low level of flexibility in the measurement definition.

The current acquisition system exploits the state-of-the-art performance of FFMM [1] and Fast Digital Integrator (FDI) [4] in order to improve the accuracy of the whole measurement and to increase the application domain. In the following, (a) the *background*, (b) the *experimental set up*, (c) the *test procedure*, (d) the *FFMM implementation*, and (e) the *experimental results* of the automatic measurement bench based on FFMM realized at CERN for the magnetic permeability measurements are presented.

9.2.1.1 Background

The principles of permeability measurements are recalled here for the particular case of tests on a ring-shaped specimen by means of a fluxmeter, when the material is subject to a particular steady state (*magnetostatic test*), or is changed from one magnetostatic condition to another.

Points on the curve of first magnetization are measured by bringing a magnetic field H to bear on the sample. Switching the field to opposite directions causes a change in the flux density B equal to twice its value. Repetition of the measurement by a gradual increase in the field will produce the set of values (B, H) determining the curve.

The value of the average magnetic induction is assessed properly by correcting for the flux contribution from the applied magnetic field outside the bulk of the sample measured by the sensing coil. For this reason, two measurements are carried out and combined, with and without the specimen inside the permeameter. Subsequently, the values of B and H can be calculated as [2]:

$$B = k_2\phi - k_3 I \tag{9.1}$$

$$H = \frac{N_1 I}{2\pi r_0}$$

with

$$k_1 = \mu_0 \frac{H}{I}$$

$$k_2 = \frac{1}{2N_2 S_a}$$

$$k_3 = k_2 \frac{\phi_0}{I} - k_1$$

$$2\pi r_0 = 2\pi \frac{r_{ext} - r_{int}}{\ln r_{ext} - \ln r_{int}}$$

where N_1 is the number of excitation windings, N_2 the number of windings of the sensing coil, $2\pi r_0$ the sample average magnetic length, r_{int} and r_{ext} the inner and outer radius of the sample, respectively, φ and φ_0 the integrated signal of the permeameter with and without sample, respectively, and S_s the sample section area. The area S_s might be difficult to assess, especially when laminated samples are used. In these cases, by knowing the density of the material and by measuring its mass, the sample volume V, and hence S_s, can be computed as $\dfrac{V}{\pi\left(r_{int} + r_{ext}\right)}$.

9.2.1.2 Experimental Set Up

In Figure 9.2a, the architecture of the automatic bench for magnetic permeability measurement based on FFMM is shown. A PC hosting the measurement application produced by FFMM is connected to a data acquisition board (*DAQ*) [5], in order to control the Voltage-Controlled (*VC*) *Power Supply* of the excitation coil of the *Split-coil Permeameter* by the analog output. The PC controls a *PXI Rack* containing (a) two FDIs configured for current acquisition and voltage integration, respectively, and (b) a board developed at CERN (Encoder Board), generating pulses used to trigger synchronously the FDIs acquisition. In Figure 9.2b, the experimental setup of the permeability measurement bench is depicted.

9.2.1.3 Test Procedure

The assessment of the magnetic permeability requires two measurements, with and without sample inside the split coils. Moreover, a preliminary demagnetization cycle is needed for bringing the sample under test into a "virgin" state. The complete measurement procedure can be summarized as follows:

1. Measurement of the average magnetic induction without the specimen to retrieve the correction factor
2. Demagnetization cycle
3. Measurement of the average magnetic induction with the specimen inside the permeameter

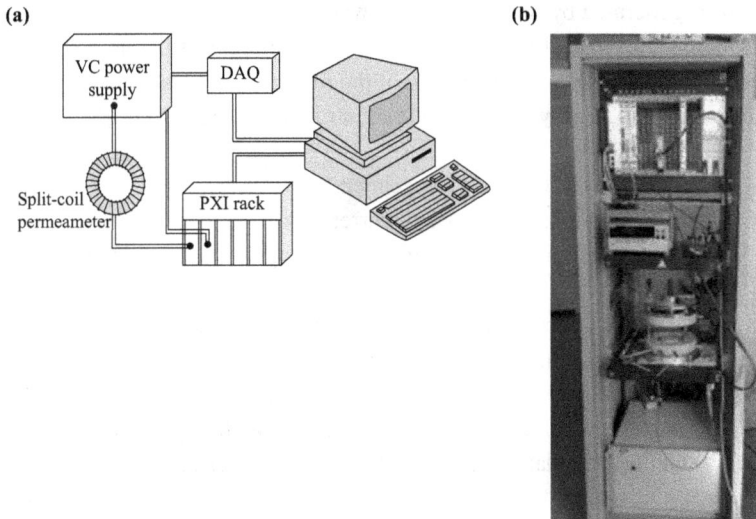

Figure 9.2. (a) Architecture and (b) experimental setup of the permeability measurement bench at CERN.

The demagnetization procedure, tuned by experimental studies, is carried out by feeding the excitation windings of the permeameter with several current plateaus. The first current plateau of 40 *A* generates a magnetization field sufficient to bring the sample into a saturation state. Subsequently, the current is decreased from 40 *A* down to 1 *mA*, with three ranges of attenuation factors according to a geometric progression. From 40 *A* down to 0.2 *A*, each plateau equals the previous divided by 1.5; then, by 1.2 down to 85 *mA*, and finally by 1.1 to 1 *mA*. Each plateau lasts 4 *s*.

Each measurement cycle, with and without sample, is carried out by powering the excitation windings through a current cycle made by increasing plateaus of opposite signs, linked by ramps of fixed slope 1.5 *A/s*. The voltage signal induced on the sensing coil is integrated to obtain the variation of the linked flux. The current is measured via a feedback signal of the VC Power Converter, synchronously with the flux by exploiting a common trigger signal.

9.2.1.4   FFMM Implementation

The software application for magnetic permeability measurement is obtained from FFMM through a formal description provided in a user script. An excerpt of the high-level user script is provided in Figure 9.3 Besides the hardware synchronization of the two FDIs by means of trigger pulses generated by the Board, a software synchronization of the devices is handled by the FFMM *Synchronizer*. In particular, suitable constructs are used to schedule the execution of the following actions without worrying about the time synchronization of parallel or series tasks:

1. Demagnetization of the specimen;
2. Start acquisition of flux and current;
3. Start generation of one cycle of the signal controlling the power converter;
4. Wait for the completion of the present current cycle;
5. Stop the acquisition of the flux;
6. Start the generation of the next current cycle and go to 3) or, if the maximum value of current is reached, stop the acquisition;
7. Convert the data obtained from the FDIs to a suitable format.

The specimen is magnetized gradually by using a current waveform consisting of a series of linear ramps and plateaus with exponentially-increasing amplitude. The current cycle referred at point 3 is composed of an initial plateau, a linear ramp with constant ramp rate, and a final plateau [3].

As an example [6], the test engineer adds a task to be executed after another one by using the statement *ADD_TASK_AFTER_TASK(first_task, second_task)*, such as shown in the case study (Figure 9.3) for the task *Set_Next_Cycle* scheduled in series to the demagnetization task.

Otherwise, if the test engineer wants to arrange the start of the execution of one task after an event produced by another task and then to let both the tasks active in parallel, he has to exploit the statement *ADD_TASK_ AFTER_EVENT(event, second_task)*, such as shown in Figure 9.3 for the task *Current_Cycle* scheduled after the event *next_cycle*. The first task to become active is the *Devices_definition* and then the *Demagnetization*. At the end of these tasks, the execution passes to the task *Set_Next_Cycle*

```
BEGIN_SCRIPT "Permeability Bench":

BEGIN_TASK "Devices_definition":
...
END_TASK

BEGIN_TASK "Demagnetization":
...
END_TASK

BEGIN_TASK " Set_Next_Cycle ":
...
END_TASK

BEGIN_TASK " Start_Acquisition ":
...
END_TASK

BEGIN_TASK " Stop_Acquisition ":
...
END_TASK

BEGIN_TASK " End_Test ":
...
END_TASK

ADD_TASK (Devices_definition)
ADD_TASK_AFTER_TASK (Demagnetization, Devices_definition)
ADD_TASK_AFTER_TASK (Set_Next_Cycle, Demagnetization)
ADD_TASK_AFTER_EVENT (Current_Cycle, next_cycle)
ADD_TASK_AFTER_EVENT (Start_Acquisition, start_cycle)
ADD_TASK_AFTER_EVENT (Stop_Acquisition, stop_cycle)
ADD_TASK_AFTER_TASK (Set_Next_Cycle, Stop_Acquisition)
ADD_TASK_AFTER_EVENT (End_test, end_measurement)

END_SCRIPT
```

Figure 9.3. MDSL script for permeability measurement.

(series execution); *Set_Next_Cycle* sets up the actual cycle of current and starts the *Current_Cycle* by throwing the event *next_cycle*. *Current_Cycle* first starts the task *Start_Acquisition* that enables the acquisition of flux and current, and then begins the generation of the actual current cycle. Finally, the test engineer stops the acquisition by throwing the event *stop_cycle*. The task *Stop_Acquisition* triggers a new execution of *Set_Next_Cycle* that restarts the loop up till the last scheduled current cycle or enables the *Data_Conversion* in order to format the output data.

9.2.1.5 Experimental Results

The on-field working of the FFMM-based automatic measurement system is highlighted by a test on a laminated soft steel sample. As a first step, the current and the flux without the sample are acquired by means of the experimental setup. In Figure 9.4, an example of the measured current cycle and the corresponding reconstructed field H are shown.

Subsequently, the test current cycle is repeated with the sample inside the permeameter. By combining the results of the two acquisitions, as previously explained, the first magnetization curve of the sample material is obtained (Figure 9.5). Finally, the sample relative permeability is estimated from the points of this curve as ratio of the fields B and H (Figure 9.6).

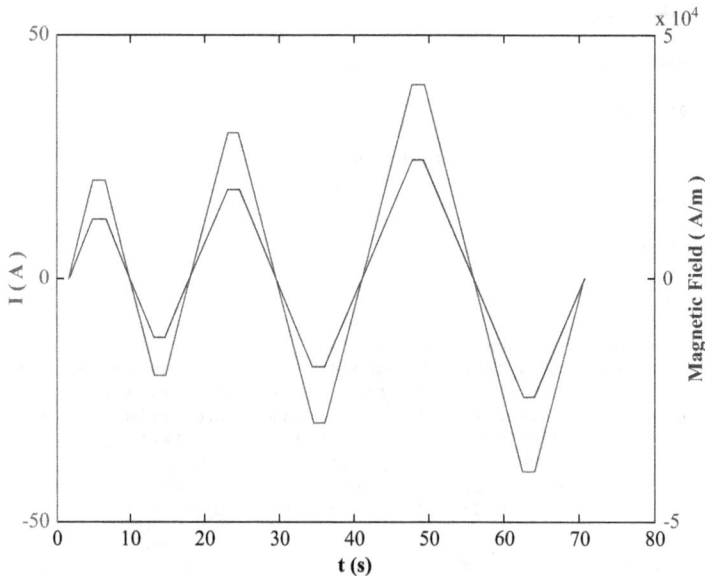

Figure 9.4. Measured current and computed magnetic field without sample.

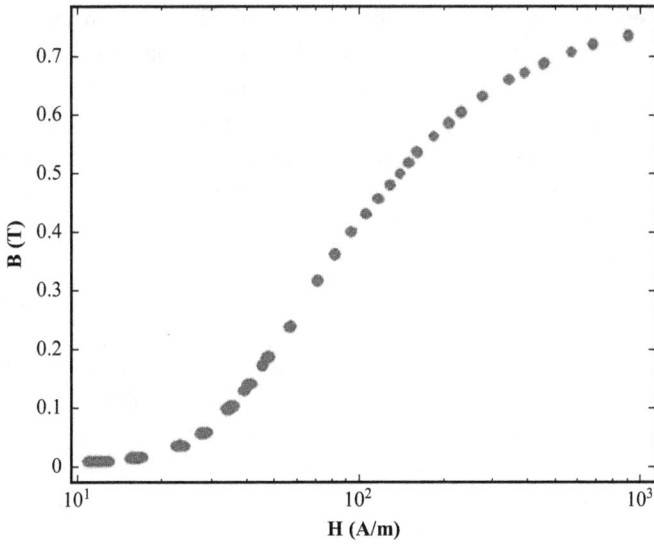

Figure 9.5. First magnetization curve of the soft steel sample.

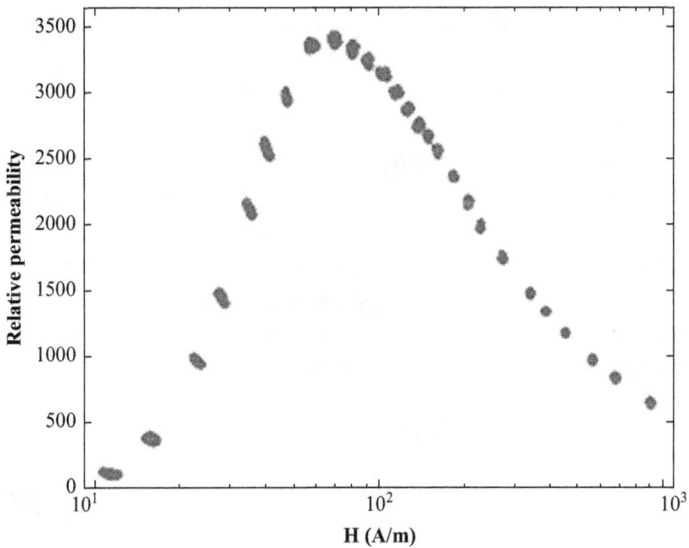

Figure 9.6. Relative permeability of the soft steel sample.

9.2.2 ROTATING COILS

FFMM has been employed at CERN for characterizing the superconducting magnets of the LHC. The LHC has unprecedented demands on
the control of the magnetic field and its distortion, during the phases of

injection, acceleration, and collision. One of the most stringent requirements during the operation of the LHC is to have a constant ratio between
dipole-quadrupole and dipole-dipole fields in order to control the variation of the *betatron tune*[1] and ensure a constant beam orbit throughout
the acceleration phase, hence avoiding particle losses. Furthermore, superconducting magnets for particle accelerators are affected by characteristic
dynamic effects, leading to field errors in the magnetic aperture of the
order of a few 10^{-4} relative to the main harmonic component. These errors,
observed and studied systematically for the first time at Tevatron [7], are
due mainly to unbalanced currents in the strands that compose the superconducting cable. In particular, LHC double-aperture 15-m long, 8.34 T
main dipole magnets are affected by a slow decay of the sextupole ($b3$)
and decapole ($b5$) components during the low-field phase of particle injection. Subsequently, as the field is ramped up for beam acceleration, these
error components "snap back" abruptly to their initial value, unbalancing
the beam orbit and giving rise to significant particle losses [8]. The tolerances on the sextupole and decapole correction are calculated from the
beam requirements [9] providing a specification for the maximum allowed
field errors. These calculations [9] yield the tolerances shown in Table 9.1
for commissioning and nominal operation phases.

Therefore, a significant measurement effort is devoted to the investigation of the dynamic field error of the main dipole magnets [10]. Fast-varying magnetic fields have been measured by designing a new measurement
station with high-speed rotating coils units and FDI [4]. A background on
this measurement method is given in Section 6.2.1 of this book.

In the following, (a) the *experimental setup*, (b) the *FFMM implementation*, and (c) the *experimental results* of the automatic measurement
bench based on the FFMM realized at CERN for characterizing the superconducting magnets of the LHC are presented.

Table 9.1. Injection harmonic tolerance (in hundreds of ppm of the main
dipole)

	Commissioning	Nominal Operation
$b3$	0.35	0.02
$b5$	-	0.1

---

[1]If a particle, along its circular path through the accelerator magnets, for any reason deviates from its nominal trajectory, it passes far from the axis of the focusing quadrupoles and oscillates around its nominal orbit (*betatron oscillation*). The
tune is the number of betatron oscillations per turn.

9.2.2.1 Experimental Setup

The architecture and experimental setup of the automatic measurement station based on rotating coils at CERN [1] are illustrated in Figures 9.7 (a) and (b), respectively. A rotating shaft, composed by 12 ceramic segments, each one holding three tangential, equal, and parallel sensing coils, equips both the apertures of the magnet (Figure 9.7b). One coil is exploited to measure the dipole field component (the so-called "absolute" signal). The connection in series opposition with a second coil provides cancellation of the dipole ("compensated" signal) for the measurement of harmonic error components with higher noise rejection [1].

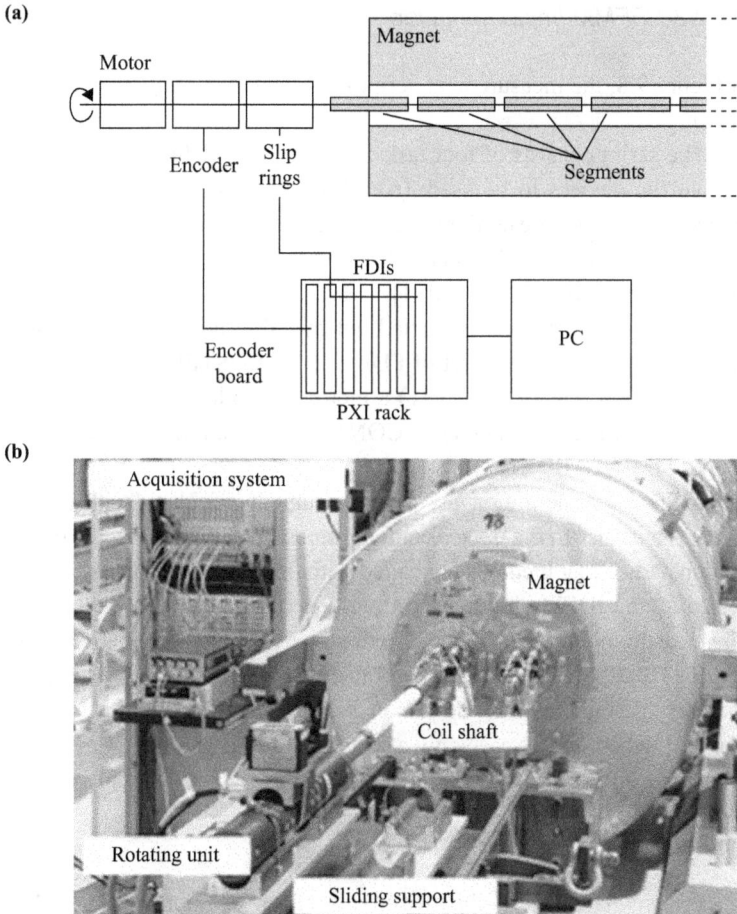

Figure 9.7. Architecture (a) and experimental setup (b) of the automatic measurement station based on rotating coils at CERN [1].

In the actual setup, the signals from consecutive segments are connected in series by three groups of four "super-segments," in order to limit the number of FDIs to six per aperture [1]. Two motors provide a rotation rate of up to 480 rpm in order to get signals at the desired time resolution. The absolute and compensated signals from each super-segment are connected to the integrator in order to measure the magnetic flux. The integrators are provided by the trigger signal from the two angular encoders, one for each motor, through a trigger board (encoder board). The magnet's supply current is read directly from the power converter's WorldFIP interface and synchronized with the acquisition trigger by means of a timing board.

9.2.2.2 FFMM Implementation

In Figure 9.8, the measurement script for superconducting magnet test is shown.

The script consists of four tasks: (a) "Devices_definition," for pointing out the devices to be used; (b) "Devices_configuration," where the device connections are configured; (c) "Devices_setting," for selecting the measurement parameters; and (d) "Measurement," where the actual measurement is started and stopped [1].

In particular, the test is carried out by using five different software devices (Figure 9.8): (a) a set of FDIs (FDI_CLUSTER), (b) an encoder board (ENCODER_BOARD), (c) a motor controller (MAXON_EPOS), (d) a power generator (POWER_CONTROLLER), and (e) and a timing board (TIMING_BOARD).

In the device definition task, each instrument is declared with a unique name as a string. In the device configuration and setting tasks, the configuration and setting methods are called with the requested parameters suggested by the IDE during the script writing (usually strings or numbers). The measurement task contains the commands related to the actions of the devices (Start, Wait, and so on) and their argument usually is the device name as a string. After the definition, each task is added to the execution tree [11], where in this case only a sequential run is needed.

9.2.2.3 Experimental Results

In Figures 9.9a and 9.9b, the measured sextuple component b_3 as a function of the current and the time, respectively, expressed in units (hundreds of ppm), namely as 10^{-4} fraction of the main field component, are

```
BEGIN_SCRIPT "Superconducting Magnet Bench":

   BEGIN_TASK "Devices_definition":
        DEF FDI_CLUSTER: "FDI_Cl_1" WITH ("FDI_Cl_1");
        DEF ENCODER_BOARD: "Enc_Brd " WITH ( "Enc_Brd" );
        DEF MAXON_EPOS: "mtr1" WITH ( "Mtr1" );
        DEF POWER_CONTROLLER: "Pow_Ctr" WITH ( "Pow_Ctr" );
        DEF TIMING_BOARD: "Time_Brd " WITH ( "Time_Brd" );
   END_TASK

   BEGIN_TASK "Devices_configuration":
        CFG FDI_CLUSTER: "FDI_Cl_1" WITH
            ([12,12,12,13,13,13],[8,9,10,8,9,10]);
        CFG ENCODER_BOARD: "Enc_Brd" WITH ( 11 , 8 );
        CFG MAXON_EPOS: "Mtr1" WITH ( 6, 100 );
        CFG POWER_CONTROLLER: "Pow_Ctr" WITH (
            "cfc",5001,50);
        CFG TIMING_BOARD: "Time_Brd" WITH ( 11,9,1000 );
   END_TASK

   BEGIN_TASK "Devices_setting":
        SET FDI_CLUSTER: SetParams   ("FDI_Cl_1",512,0.1,20);
        SET ENCODER_BOARD: SetParams ("Enc_Brd", 1, 512, 1,
            0);
        SET MAXON_EPOS: SetParams("Mtr1", 0, 10 , 10 , 60);
        SET TIMING_BOARD: TimeStampTrigger_DF1("Time_Brd");
   END_TASK

   BEGIN_TASK "Measurement":
        CMD MAXON_EPOS: Start ("Mtr1");
        CMD MAXON_EPOS: Wait ("Mtr1");
        CMD POWER_CONTROLLER: Startlog("Pow_Ctr",
            "current.txt");
        CMD FDI_CLUSTER: StartAcquisition ("FDI_Cl_1",
            "dfluxes1.txt");
        CMD ENCODER_BOARD:StartTrigger ("Enc_Brd", 1 );
        CMD TIMING_BOARD: ReadTimeStamp("Time_Brd",1);
        CMD TIMING_BOARD: DisableTimeStamp("Time_Brd" );
        WAIT 1000 ms;
        CMD FDI_CLUSTER: StopAcquisition ("FDI_Cl_1");
        CMD FDI_CLUSTER: WaitAcquisition ("FDI_Cl_1");
        CMD ENCODER_BOARD: Stop_Trigger ("Enc_Brd", 1 );
        CMD MAXON_EPOS: Stop("Mtr1");
        CMD MAXON_EPOS: WaitStop ("Mtr1");
        CMD POWER_CONTROLLER: Stoplog("Pow_Ctr");
   END_TASK

   ADD_TASK (Devices_definition)
   ADD_TASK_AFTER_TASK
   (Devices_configuration, Devices_definition)
   ADD_TASK_AFTER_TASK (Devices_definition,
   Devices_setting)
   ADD_TASK_AFTER_TASK (Devices_setting, Measurement)

END_SCRIPT
```

Figure 9.8. Superconducting magnet test script.

(a)

(b)

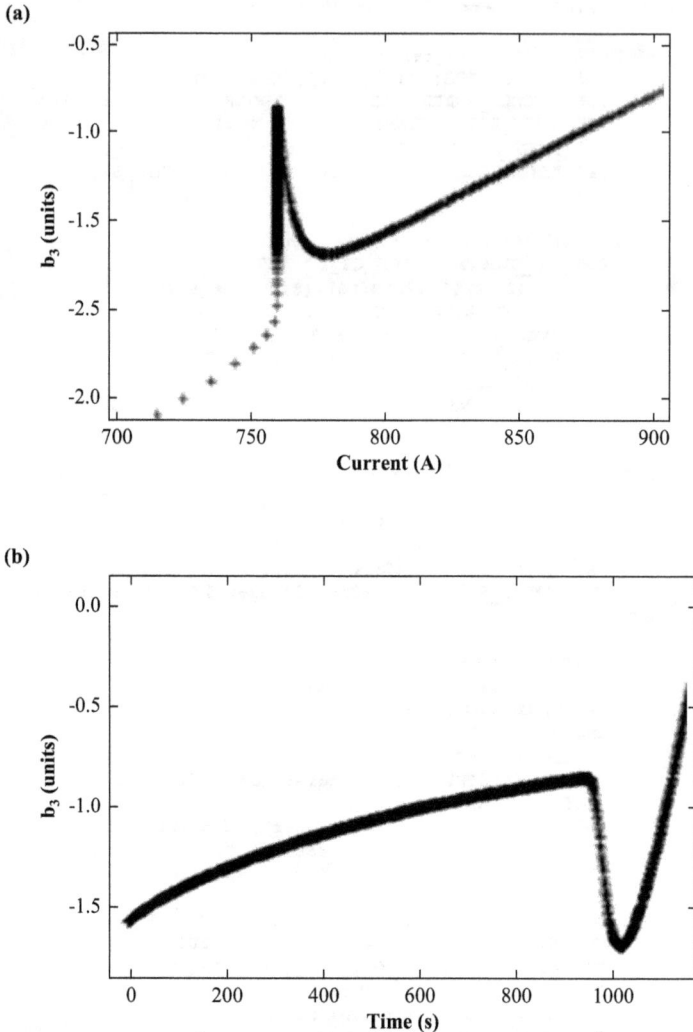

Figure 9.9. Measured sextuple component b_3 versus (a) current and (b) time, in units (10^{-4} fraction of the main field component).

highlighted. The "decay" and "snapback" phenomena [12], typical in superconducting magnets, are visible clearly [1].

9.2.3 LHC TRACKING TEST

A further specific test (*tracking test*) has been performed on the LHC cryo-magnets with twofold purposes:

1. Generate the current ramps for the main superconducting magnets producing the expected magnetic fields.
2. Generate the current ramps to supply the corrector magnets and compensate the sextupole and decapole field errors in the main dipole.

Next, (a) the *background*, (b) the *experimental setup*, (c) the *test procedure*, and (d) the *experimental results* of the tracking test are presented.

9.2.3.1 Background

The decay amplitude is affected by the power history of the magnet, and particularly by the precycle flat top current and duration. The system Field Description for the LHC (*FiDeL*) [13], modeling the field variations during injection, acceleration, and collision, was developed to cope with the differences between the current cycles during the tests and the expected cycles in the actual machine operation. *FiDel* is a feed-forward system used to forecast and compensate the field variations within the commissioning tolerance, in order to bring the beam to its nominal parameters by means of suitable controls.[2] In practice, the LHC ring is divided into eight sectors, where the compensation actions are actuated independently by means of power converters supplying the series of correctors. An average parametric field model for each sector is therefore necessary. The model is based on the identification and decomposition of the effects that contribute to the total field in the LHC dipoles. Each effect is modeled theoretically or empirically. The parameters of the model are obtained from a synthesis of the information available from magnetic field measurements in warm and cold conditions, in particular from the measurements of (a) all the magnets at room temperature and (b) one fifth of the magnets at cryogenic temperature ($1.9\ K$).

In this section, the application of FFMM to the rotating coil technique devoted to the field harmonic correction of the tracking test is presented. This procedure strongly relies on field harmonic analysis, for which rotating coils are one of the most accurate techniques. In particular, in the following, the compensation of the sextupole term for the commissioning phase is reported.

---

[2]For the LHC, a system solely based on beam feedback may be too demanding. The LHC beam control therefore requires a forecast of the magnetic field and the multipole field errors to reduce the burden on the beam-based feedback.

Past measurement campaigns [9] highlighted the need to improve the harmonic compensation of the third-harmonic (b_3) component of the main LHC dipoles. In particular, measurements had already been carried out by means of the standard measurement equipment [14], but the time resolution obtained in the field estimation was intrinsically limited by the acquisition hardware, so that a new harmonic value was available only every 20 s. A new, fast hardware was developed to overcome this limitation [15]. Currently the harmonic estimation can be updated at a rate of 8 S/s. Anyway, the new hardware is still needed for proper acquisition and control software. This software was realized by means of the FFMM.

The sextupole harmonic component in the LHC dipole is compensated by applying to the magnets a current $I(t)$ computed from (a) the integral $B3$ and (b) the sextupole corrector Transfer Function (TF). TF (expressed in $T\,m/A$) is obtained by averaging the ratio of the measured fields and currents of the correctors on the two apertures of the dipole:

$$TF = \frac{L}{2}\left\{ mean_t \left[\frac{B_3^{MCS1}(t)}{I(t)} \right] + mean_t \left[\frac{B_3^{MCS2}(t)}{I(t)} \right] \right\} \qquad (9.2)$$

where L is the length of the super segment used for sextupole magnets measurement (Figure 9.7a). The integral B_3 (expressed in $T\,m$) is defined as:

$$\bar{B}_3(t) = \sum_{i=1}^{3} L_i B_{i,3}(t) \qquad (9.3)$$

where L_i and $B_{i,3}(t)$ are the effective length and the measured third harmonic of the i-th super segment, respectively. The effective length is adjusted to take into account the contribution of the gaps between the coils. All harmonic field values are expressed in T measured at a given reference radius, conventionally 17 mm for the LHC.

Finally, by combining Equation 9.2 and Equation 9.3, the current for dipole harmonic compensation turns out to be:

$$I_{MCS}(t) = -\frac{\bar{B}_3(t)}{TF} \qquad (9.4)$$

The average value of the TF (9.2) of 95 T/A is consistent with the previous measurements of the sextupoles installed in the machine [9], and the RMS fit error of 3 $\mu T/A$ proves a satisfying TF linearity.

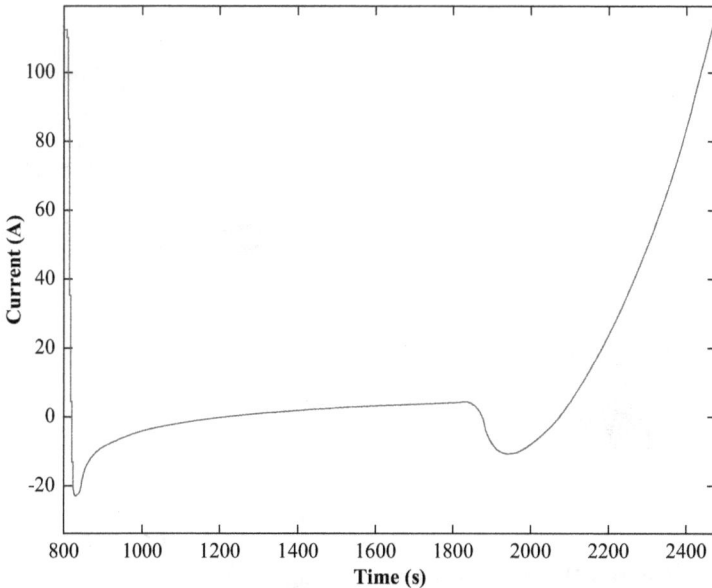

Figure 9.10. Computed MSCs powering current cycle for sextupole compensation.

In Figure 9.10, the current curve for the corrector magnets computed through (9.4) is shown.

9.2.3.2 Experimental Setup

In Figure 9.11, the architecture of the measurement system is shown. The core is the fast equipment for harmonic coils measurements. Varying magnetic fields are measured by designing the measurement station according to the main specification of improving the bandwidth. This is achieved by means of high-speed rotating units and associated electronics. The algorithm for harmonic resolution enhancement presented in Section 6.2.1 was not employed, because the current limitations of the power converter control system makes it pointless. Both the apertures of the cryo-assembly are equipped with a rotating shaft, made of 12 pivoting ceramic segments each holding three tangential, equal, and parallel pick-up coils [16].

One coil is normally used to measure the dipole field component (the so-called "absolute" signal), while the connection in series opposition with a second coil provides cancelation of the dipole ("compensated" signal) and ensures higher noise rejection for the measurement of harmonic error components. The shaft covers the whole length of the LHC dipole and the last segment captures the sextupole corrector field in its entirety. In this configuration, such as already explained, the signals from consecutive

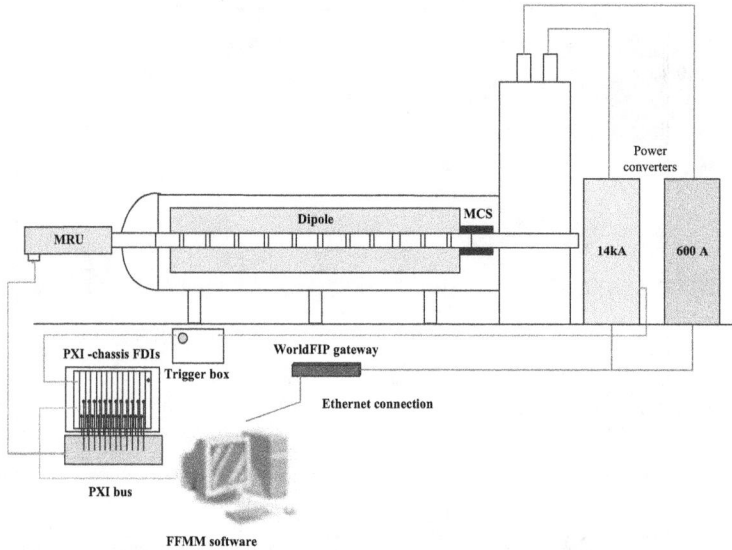

Figure 9.11. Architecture of the tracking test measurement station.

segments are connected in series by three groups of four segments, constituting three "super segments" with the purpose of limiting the number of necessary integrators to six per aperture. Two *Micro Rotating Units* [15] provide a rotation speed of up to 8 *rps* in order to get voltage signals with the desired time resolution. The absolute and compensated signals, from each super segment, are the input of a FDI [17] measuring the magnetic flux linked with the super segment coils. The pulses from the angular encoders trigger the integration time of the FDIs and the acquisition of the supply magnet current: The synchronization between the magnetic flux sample and current measurement is thus ensured. The software used to handle the station and to retrieve the current reading, via Ethernet connection from the power supply controller, is obtained through the FFMM.

9.2.3.3 Test Procedure

The test procedure for the harmonic compensation is composed of the following steps:

1. Measurement of the integral dipole field and error components during a nominal LHC machine cycle;
2. Measurement of the transfer function of the two superconducting sextupole corrector magnets (MCS) installed in line with each dipole aperture in the same cryo-assembly [18];

3. Computation of the compensation current for the MCS from the results of points 1 and 2;

4. Measurement of the integral field when the main dipole performs an LHC cycle and the MCS are supplied with the compensation current, and estimation of the residual field errors.

The first step is aimed at characterizing the LHC dipole magnet during a nominal machine cycle (LHC cycle, Figure 9.12) [19].

In particular, the measurement is carried out to get a reference behavior, without compensation, of the integral harmonic component B_3. The nominal LHC cycle has a ramp-up at 10 A/s from 350 A to an injection current plateau at 760 A, lasting about 1000 s, to simulate the particle injection at constant field (Figure 9.12). This is followed by a Parabolic Exponential Linear Parabolic (PELP) [19] profile, a 1000 s flat top at a nominal current of 11,850 A, and a ramp-down at 10 A/s to the minimum current of 350 A.

The LHC cycle is preceded by a precycle aimed at bringing the magnet into a reproducible magnetic state. The repeatability of the sextupole obtained in the characterization phase of the magnet is in the order of 10^{-4} *units*. Regardless, without well-defined cycling procedures, the reproducibility of the sextupole during LHC cycles is ~ 0.1 *units* [9]. For obtaining the target value of 0.02 *units* in the compensation of b_3 (Table 6.1), suitable cycling procedures are therefore defined.

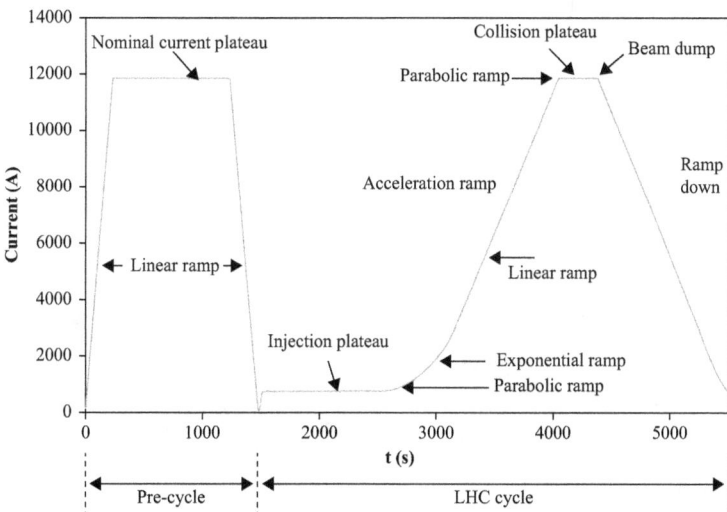

Figure 9.12. LHC standard current cycle.

The second step is aimed at (a) computing the TF, that is, the ratio between the field and the current, for the two sextupole correctors, and (b) verifying the linearity of such a TF. The resulting TF allows the required sextupole excitation current to be computed for the compensation of the B_3 field inside the dipole. The field is measured during several ramp cycles, that is, from 0 A up to the nominal current of 550 A, down to -550 A and back to 0 A with a ramp rate of ± 10 A/s.

The third step is the main measurement procedure. The sextupole is fed with the current curve computed via the TF and at the same time the LHC dipole is fed with the nominal LHC cycle, both cycles being tightly synchronized (<1 ms). The results of such a measurement highlight the quality of compensation for the third harmonic in the dipole.

In Figure 9.13, a domain specific language (DSL) user script for rotating coil-based measurements is provided. The script contains two

```
BEGIN_SCRIPT main_LHC_cycle:

/*******************************/
// Device Definition
/*******************************/
DEF FDI_CLUSTER:      FDI_CLuster1      WITH (numberOf_FDI);
DEF ENCODER_BOARD: Enc_B1               WITH ("1", "1", "CERN");
DEF MAXON_EPOS:      Mot_C1             WITH ("1","1","1",15);
DEF POWER_CONTROLLER: pow               WITH (17,"cfc-sm18-r2",1905,"*******","*******");
DEF POWER_CONTROLLER: pow2              WITH (17,"cfc-sm18-r2",1905,"*******","*******");

/*******************************/
// Device Configuration
/*******************************/
CFG FDI_CLUSTER:      FDI_CLuster1      WITH (Cluster_bus1, Cluster_slot1);
CFG ENCODER_BOARD: Enc_B1               WITH (Encoder_bus, Encoder_slot);
CFG MAXON_EPOS:      Mot_C1             WITH (4, Motor_timeout);

/*******************************/
// Measurement Task Definition
/*******************************/
//--------------------------------
BEGIN_MTASK Device_Setting:

    SET FDI_CLUSTER:      Params2 (FDI_CLuster1, spt, SamplePerTurn, Cluster_abs_gain_, Cluster_comp_gain_, 2, 500000, spt2,
60);
    SET FDI_CLUSTER:      Stop_Source ( FDI_Cluster_1, surceStop);
    SET MAXON_EPOS:      OperationMode ( Mot_C1, 0);
    SET MAXON_EPOS:      Acceleration ( Mot_C1, Motor_Acc_);
    SET MAXON_EPOS:      Deceleration ( Mot_C1, Motor_Dec_);
    SET MAXON_EPOS:      Velocity ( Mot_C1, Motor_Vel_);
END_MTASK

//--------------------------------
BEGIN_MTASK Flux_Measurement:
//--------------------------------
    PRINT "To start the measurement of LHC cycle digit yes";
    SET ENCODER_BOARD:            Enc_B1 (Encoder_Channel, Number_of_Angular_Encoder_Pulse, 1, divF, ndivF, 0);
    PRINT "START LHC CYCLE MEASUREMENT";
    PRINT "start motors" ;
    CMD MAXON_EPOS:      Start (Mot_C1);
    SET MAXON_EPOS:      VelocityThreshold (Mot_C1, Motor_Vel_);
    CMD MAXON_EPOS:      WaitVelocityThreshold (Mot_C1);
    CMD FDI_CLUSTER:      Acquisition (FDI_Cluster_1, Encoder_Channel );
    CMD FDI_CLUSTER:      Wait_Acquisition (FDI_CLuster_1);
    CMD FDI_CLUSTER:      Stop_Acquisition (FDI_Cluster_1 );
    CMD ENCODER_BOARD:            Stop_Encoder_Trigger ( Enc_B1, Encoder_Channel );
    PRINT "stop motor";
    CMD MAXON_EPOS:      Stop ( Mot_C1);
    CMD MAXON_EPOS:      WaitStop ( Mot_C1);
    PRINT "End LHC measurement cycle";
    PRINT "data conversion";
END_MTASK

ADD_TASK Device_Setting;
ADD_TASK_AFTER_TASK Device_setting Flux_Measurement;

END_SCRIPT
```

Figure 9.13. DSL script for rotating coil-based measurement.

measurement tasks handling devices setting and flux measurement procedure. The FFMM *Synchronizer* is in charge for scheduling their execution
in sequence.

The desired sextupole excitation cycle is to be approximated by interpolating the current curve of Figure 9.12 with few linear segments, owing
to limitations of the power converter control system (to be updated in the
near future). The harmonic measurements are performed at 1 *S/s*, that is,
with a coil rotation speed of 1 *rps*, which is still well below the theoretical bandwidth of the instrumentation, albeit 20 times faster than with the
instrumentation available during the series tests of the LHC magnets.

9.2.3.4 Experimental Results

The preliminary results reported in this section focuses on highlighting
the capability of the new setup of attaining at the first iteration a compensation level of the integral sextupole harmonic very close to the previous
campaign [9].

In Figure 9.14, the integral b_3 in units, measured during the reference measurement of the LHC cycle in the aperture 1 of the dipole
MB2425, is shown. In the same figure, the integral b_3 during the harmonic

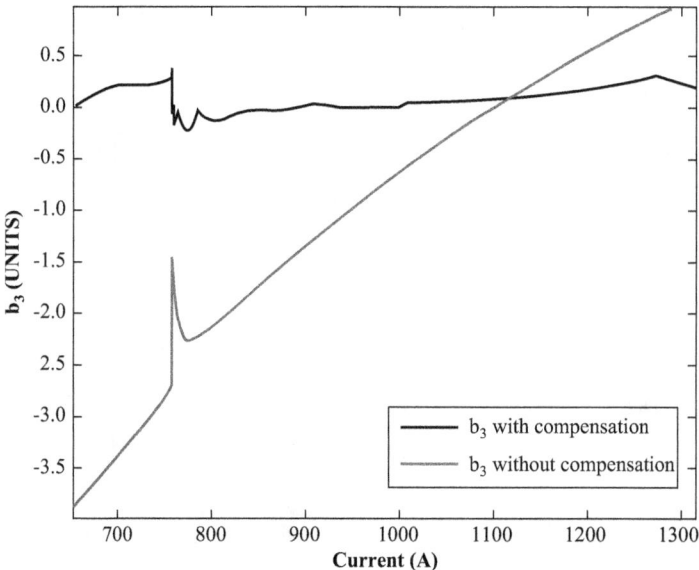

Figure 9.14. Integral b_3 component versus current with and without compensation, in the dipole magnet *MB2524* during an LHC cycle.

compensation measurement, with the linear interpolated current supplying the MCSs, is reported. A more detailed view of the residual b_3 with compensation is provided in Figure 9.15. A tolerance of 0.6 *units*, roughly corresponding to a reduction by a factor of 3 of the snapback swing, is achieved already at the first iteration. The decay and snapback transient is detected with unprecedented detail, in particular, considering the amplitude of the peak [15].

The increased resolution is highlighted in the comparison with the results of the standard measurement equipment shown in Figure 9.16, for an acquisition on a single segment with the new acquisition system running at maximum speed (8 *rps*).

9.3 FLEXIBILITY EXPERIMENTAL TESTS

This section deals with the flexibility test of software frameworks for measurement applications, and, in particular, of the FFMM. After prototyping, experimental applications, and analysis of code quality, a flexibility characterization of the framework is needed. As part of the wider scenario aimed at the characterization of the framework started in Chapter 5, the

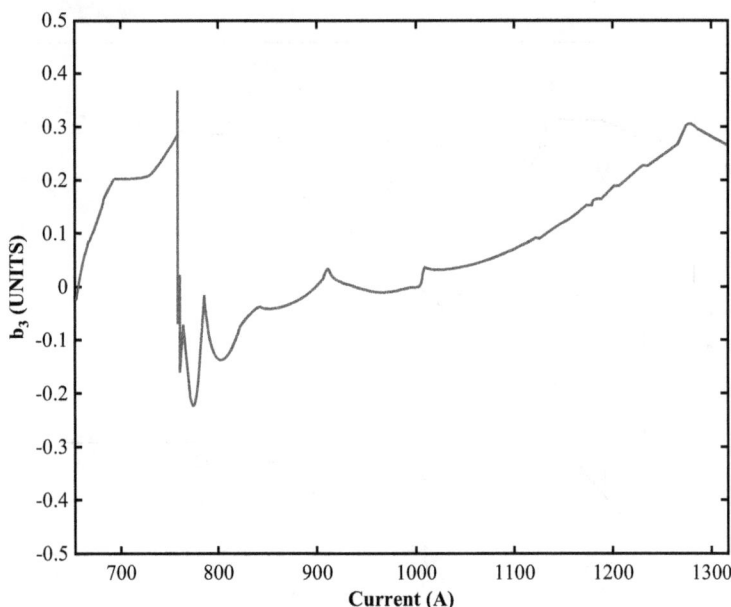

Figure 9.15. Residual integral b_3 component versus current with compensation, in the dipole magnet *MB2524* during an LHC cycle.

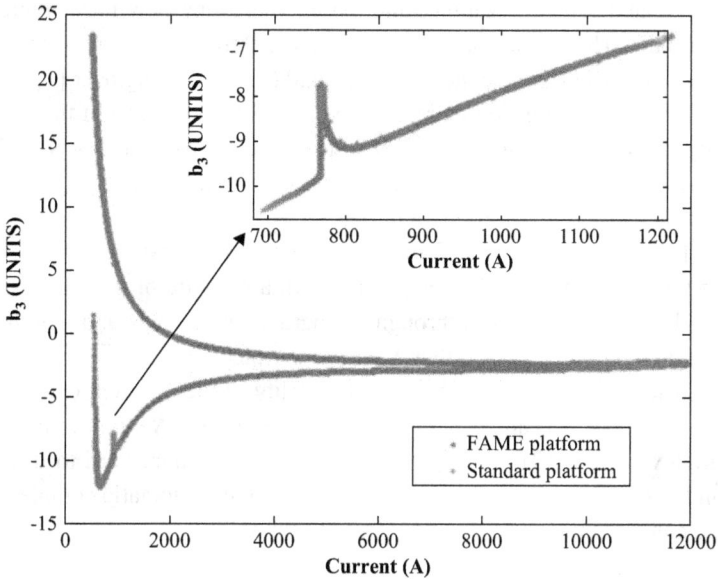

Figure 9.16. Estimation of the sextupole with the old standard and the FFMM platform (FAME).

twofold purposes of this section is (a) to introduce specific metrics suitable for assessing the degree of flexibility achieved by a software framework for measurement applications and (b) to present experimental results for some typical application scenarios from the CERN release 3.0 of FFMM.

9.3.1 EXPERIMENTAL RESULTS

The FFMM at CERN was designed in order to satisfy the requirements for a wide range of magnetic measurement applications, thus the most probable scenarios to deal with are the different techniques currently used for testing magnets for accelerators, besides those that will be developed in the future.

The framework is based on Object-Oriented and Aspect-Oriented Programming (OOP and AOP), therefore the modules involved in these scenarios are *methods*, *classes*, and *aspects*. In a preliminary analysis phase, the classes of changes due to the different measurement techniques were classified (by increasing flexibility) as: (1) adding or modifying software modules implementing the devices, (2) changing the strategies for handling the services provided by the framework (e.g., fault detection, logging, and synchronization), and (3) implementing new measurement algorithms. The classes of changes involve different users of the framework, namely (1) and (2) the developer, and (3) the test engineer [1].

In the following, some preliminary experimental results of the flexibility assessment are illustrated. The tests were carried out at CERN on the release 3.0 of FFMM for different measurement methods. The experimental results are summarized in Table 9.2.

The generalized evolution cost metric is obtained by fixing $\mu = CYCLO$ (Cyclomatic Complexity [20], a measure of the number of linearly independent paths through a program source code and therefore of its logical complexity), thus yielding the metric $C_{Modules}^{CYCLO}$. This metric is used to compare the degree of flexibility of the different classes of changes, and not as an absolute measure of flexibility. A high cyclomatic complexity (>10 [21–24]) denotes a complex procedure hard to understand, test, and maintain. Therefore, the lower the cyclomatic complexity (and consequently $C_{Modules}^{CYCLO}$), the higher the flexibility.

9.3.1.1 Adding or Modifying a Device

When new devices are required by a measurement application, the effort for their implementation cannot be avoided completely. In this case, the flexibility is, as a consequence, limited intrinsically [1]. Nevertheless, FFMM is fairly flexible toward this class of changes, because it helps the user effectively in developing the related new components. Namely, it provides services, such as fault detection and event handling, whose infrastructure is accessible easily and whose implementation is customizable with limited effort.

Table 9.2. Generalized evolutions cost metric for different classes of changes in FFMM

	Class of change	User involved	$C_{Modules}^{CYCLO}$
↓ Increasing flexibility	Add device	Developer	$\propto$ #methods,events,faults
	Change device interface	Developer	$\propto$ depth inheritance hierarchy
	Change fault detection strategy	Developer	$\propto$ #faultsinvolved
	Change measurement procedure	Test engineer	0

The possible changes at device level can be classified as (a) adding the device into the framework from scratch and (b) modifying it to satisfy new requirement when it already exists, by adapting some method implementation or its interface. The cost of adding a new device strongly depends on its size. Formally, rather than through the lines of code (LOC), this cost can be expressed as the sum of the cyclomatic complexity of all its methods, including additional code devoted to faults and events handling [1]. It is computed as the average cyclomatic complexity of a software unit (method) multiplied by the number of units implemented. The generalized evolution cost metric results therefore proportional to the number of member functions, events, and faults of the new class (Table 9.2). The member functions of the class are likely to be more complex than the methods handling the events and faults, thus the generalized evolution cost metric usually depends more on the former set of functions. If a class interface has to be modified, for example, by adding or removing a method, the change will involve many modules because typically a device is part of a hierarchy of classes in a generalization relationship. The effort to add or remove a method is fixed and determined by its own complexity, thus the growth of the evolution cost metric depends only on the depth of the inheritance hierarchy (Table 9.2). In the design phase of FFMM, the maximum depth was kept to a reasonable value, thus this class of changes requires a limited effort. The evolution cost estimation strongly depends on the device considered.

In order to provide a quantitative example, in the following the driver for the device Encoder Board, developed at CERN and employed in different scenarios typical of the magnetic measurements, is taken into account. The device is part of the hierarchy of classes. Adding the device requires a considerable programming effort, anyway FFMM provides support in the following ways: Libraries implementing communication features on different buses are supplied, so that all the required functionalities are already available and accessible through a suitable interface. Furthermore, FFMM already implements and makes available infrastructures for event handling and fault detection. The tasks of exploiting events and improving system fault tolerance are therefore extremely simplified for the user. He just needs to add few small modules to extend the event structure and the fault detection logic. The generalized evolution cost metric, computed as the sum of the total cyclomatic complexity of the modules to be added, has a value of 301 in the particular case considered. This value is provided just as an example, since it strongly depends on the evolution step under analysis. Anyway, to give insight on its meaning, it can be observed that in FFMM an average device has 35 member functions, and an average device member function has $CYCLO = 2$. Therefore, since the complexity of a device lies mainly in its member functions, on average adding a new device costs $35 \times 2 = 70$. With respect to this value, the Encoder Board is found to be significantly above the average complexity level.

9.3.1.2 Changing Service Strategies

FFMM provides many services to help the user in employing the framework and enlarging its application domain. The choice of OOP reduces the number of modules affected by possible changes, thus assuring a good level of flexibility. Moreover, some services (Chapters 4 and 7) were implemented by means of AOP. As an example, here only the fault detection is considered, because it is a fundamental part of all the devices (Section 4.3). By this solution, the classes of FFMM are oblivious of triggering the execution of a specific code in the related aspects providing the services. Classes and aspects are therefore completely decoupled, further increasing software flexibility. Namely, a change in the fault detection strategy typically involves only one module, without affecting in any way the corresponding device. The complexity of such a change can be estimated as the average complexity of a fault handling method multiplied by the number of methods to be modified, and is therefore proportional to the number of faults involved in the change (Table 9.2).

To provide an assessment, for the fault detection code specific of the Encoder Board, a $C_{Modules}^{CYCLO} = 4$ is achieved, while for fault detection code common to other devices, a $C_{Modules}^{CYCLO} = 25$, for a total generalized evolution cost of 29.

9.3.1.3 Implementing New Measurement Algorithms

Several measurement techniques are currently employed for testing accelerator magnets, such as fixed and rotating coils, as well as stretched wire [25]. FFMM was designed to reduce drastically the amount of code affected by modifying the measurement procedure.

The test engineer interacts with the framework mainly through the User Script, which is a formal description of the measurement procedure. All the changes required by a new measurement algorithm are focused in the User Script, without affecting any other modules. In this case, the framework provides the highest degree of flexibility, with $C_{Modules}^{CYCLO} = 0$ (Table 9.2). This result was proven experimentally by developing the application for permeability measurements described in Section 9.2.1 by means of devices already developed and previously employed for the rotating coil benches. The system for permeability measurement was developed at CERN in the 1960s, and since then has been used by means of a semiautomatic test station [26]. It is therefore remarkable the possibility to develop quickly from scratch the required control and acquisition software through FFMM.

9.4 DISCUSSION

In this chapter, the FFMM was shown to be employed successfully in the field to produce the software applications required by current needs at the CERN test facilities. In particular, the framework proved its effectiveness in developing software for measurements with very different requirements, based both on rotating and fixed coils. The former technique was employed for the estimation and compensation of the sextupolar component of the field generated by a superconducting LHC dipole. The latter technique was used to estimate the permeability of a soft steel sample.

The results of harmonic compensation and permeability measurement were presented, highlighting the resolution improvement attained in the harmonic estimation and the capability of FFMM of producing quickly, and with a limited effort, the acquisition and control software for both applications, in particular, for the split-coil permeameter, for which an automatic test station had never been developed before

An experimental approach to the software's flexibility assessment of measurement frameworks is shown. In particular, this approach is applied in the context of the FFMM at CERN. FFMM was designed to be flexible, reusable, maintainable, and portable. A complete release of FFMM is available, and its effectiveness on the field has already been proven in Chapter 6, thus the evaluation of its degree of flexibility completes the more comprehensive phase of software quality assessment, aimed at stating the fulfillment of the challenging project goals.

The flexibility of the system cannot be stated in absolute terms, but only with respect to specified classes of changes, involving different users. The results highlight that the framework achieves increasing degrees of flexibility moving from the programming level to the user script level, and at the same time from the point of view of the developer to that of the test engineer. The highest flexibility is attained for the changes involving the measurement procedure, namely at the level where flexibility was mainly required.

REFERENCES

[1] Arpaia, P., M. Buzio, L. Fiscarelli, and V. Inglese. November 2012. "A Software Framework for Developing Measurement Applications Under Variable Requirements." *AIP Review of Scientific Instruments* 83, no. 11. doi: 10.1063/1.4764664

[2] Dieterly, D.C., R.F. Edgar, A.H. Fredrick, J.W. Hale, D.H. Jones, H.W. Lamson, W.T. Mitchell, R.E. Mundy, C.D. Owens, A.C. Beiler, I.L. Cooter, W.S. Eberly. 1970. *Direct-Current Magnetic Measurements for Soft Magnetic Materials. Baltimore, Md:* American Society for Testing and Materials.

[3] Arpaia, P., M. Buzio, L. Fiscarelli, G. Montenero, and L.Walckiers. May 3–6, 2010. "High Performance Permeability Measurement: A Case Study at CERN." In *Proceeding International Instrumentation and Measurement Technology Conference*, pp. 58–61. Austin, TX: IEEE.

[4] Arpaia, P., L. Bottura, L. Fiscarelli, and L. Walckiers. 2012. "Performance of a Fast Digital Integrator in On-Field Magnetic Measurements for Particle Accelerators." *Review of Scientific Instruments* 83, no. 2, 024702. doi: http://dx.doi.org/10.1063/1.3673000

[5] National Instruments. 2014. *NI M Series Multifunction DAQ for PCI, PXI, and USB*, http://sine.ni.com/nips/cds/view/p/lang/en/nid/14114

[6] Arpaia, P., L. Fiscarelli, G. La Commara, and C. Petrone. 2011. "A Model-Driven Domain-Specific Scripting Language for Measurement System Frameworks." *IEEE Transactions on Instrumentation and Measurement* 60, no. 12, pp. 3756–66. doi: 10.1109/TIM.2011.2149310

[7] Finley, D.A., D.A. Edwards, R.W. Banft, R. Johnson, A.D. MC Inturff, and J. Strait. 1987. "Time Dependent Chromaticity Changes in the Tevatron." *Presented at the 12th Particle Accelerator Conference.* Batavia, IL: Fermilab.

[8] Ambrosio, G., P. Bauer, L. Bottura, M. Haverkamp, T. Pieloni, S. Sanfilippo, and G. Velev. June 2005. "A Scaling Law for the Snapback in Superconducting Accelerator Magnets." *IEEE Transaction on Applied Superconductivity* 15, no. 2, pp. 1217–20. doi: http://dx.doi.org/10.1109/tasc.2005.849535

[9] Xydi, P., N. Smmut, R.A. Fernandez, L. Bottura, G. Deferne, M. Lamont, J. Miles, S. Sanfilippo, M. Strrzelczy, and W. Delsolaro. 2008. "A Demonstration Experiment for the Main Field Tracking and the Sextupole and Decapole Compensation in the LHC Main Magnets." In *LHC Project Report* 1083, CERN. Geneva, Switzerland.

[10] Bottura, L. April 11–17, 1997. "Field Dynamics in Superconducting Magnets for Particle Accelerators." *CERN Accelerator School on Measurement and Alignment of Accelerator and Detector Magnets*, pp.79–105. Anacapri, Italy: CERN.

[11] Fowler, M., and R. Parsons. 2010. *Domain-Specific Languages.*: Addison-Wesley. Westford, MA, USA.

[12] Bottura, L., L. Walckiers, and R. Wolf. June 1997. "Field Errors Decay and 'Snapback' in LHC Model Dipoles." *IEEE Transactions on Applied Superconductivity* 7, no. 2, pp. 602–5. doi: http://dx.doi.org/10.1109/77.614576.

[13] Sammut, N., L. Bottura, and J. Micallef. 2006. "A Mathematical Formulation to Predict the Harmonics of the Superconducting LHC Magnets." *Physicals Review Special Topics–Accelerators and Beams* 9, no 1, p. 012402. doi: http://dx.doi.org/10.1103/physrevstab.9.012402

[14] Billan, J., L. Bottura, M. Buzio, G. D'Angelo, G. Deferne, O. Dunkel, P. Legrand, A. Rijllart, A. Siemko, P. Sievers, S. Schloss, and L. Walckiers. 2000. "Twin Rotating Coils for Cold Magnetic Measurements of 15 m Long LHC Dipoles." *IEEE Transactions on Applied Superconductivity* 10, no. 1, pp. 1422–26. http://dx.doi.org/10.1109/77.828506

[15] Brooks, N.R., L. Bottura, J.G. Perez, O. Dunkel, and L. Walckiers. June 2008. "Estimation of Mechanical Vibrations of the LHC Fast Magnetic Measurement System." *IEEE Transactions on Applied Superconductivity* 18, no. 2, pp. 1617–20. doi: http://dx.doi.org/10.1109/tasc.2008.921296

[16] Arpaia, P., M. Buzio, L. Fiscarelli, G. Montenero, J.G Perez, and L. Walckiers. May 3–6, 2010. "Compensation of Third-Harmonic Field Error in LHC Main Dipole Magnets." In *Proceedings of the IEEE Instrumentation and Measurement Technology Conference*. Austin, TX: IEEE.

[17] Arpaia, P., A. Masi, and G. Spiezia. April 2007. "Digital Integrator for Fast Accurate Measurement of Magnetic Flux by Rotating Coils." *Instrumentation and Measurement, IEEE Transactions* 56, no. 2, pp. 216–20. doi: http://dx.doi.org/10.1109/tim.2007.890787

[18] CERN. 2004. *LHC Design Report*. Geneva, Switzerland: CERN.

[19] Sanfilippo, S., L. Bottura, M. Buzio, and E. Effinger. 2002. "Magnetic Measurements for 15-M Long Dipoles–Extended Program of Tests." Internal note LHC-MTA-IN-2002-183

[20] McCabe, T.J. December 1976. "A Complexity Measure." *IEEE Transactions on Software Engineering* 2, no. 4, pp. 308–20. doi: http://dx.doi.org/10.1109/tse.1976.233837.

[21] Lanza, M., and R. Marinescu. 2006. *Object-Oriented Metrics in Practice*. Secaucus, NJ, USA: Springer-Verlag New York, Inc.

[22] French, V.A. September 1999. "Establishing Software Metric Thresholds." In *WSM 99: International Workshop on Software Measurement*, pp. 43–50. Lac Superieur, Canada: IEEE.

[23] Sbavaglia, R., 2005. "Le metriche e il loro utilizzo nello sviluppo del software" (in Italian). http://torlone.dia.uniroma3.it/sistelab/annipassati/sbavaglia.pdf

[24] Nikora, A.P. et al. 2002. "Software Metrics in Use at JPL Applications and Research" http://trs-new.jpl.nasa.gov/dspace/bitstream/2014/8671/1/02-1209.pdf.

[25] Henrichsen, K.N. April 1997. "Overview of Magnet Measurement Methods." *CERN Accelerator School on Measurement and Alignment of Accelerator and Detector Magnets*, pp.79–105. Anacapri, Italy: CERN.

[26] Henrichsen, K.N. 1967. "Permeameter." In *Proceeding Second International Conference on Magnet Technology*. pp. 735–9. Oxford, U.K: Rutherford Laboratory.

INDEX

A

abstraction, 50
ActivATE platform, 11
Active device, 175–176
Advanced User Interface
    Generator
  configuration, 160–163
  discussion, 249
  input/output values, 163–164
  magnetic permeability
    measurement, 247–249
  plotting functions, 164–165
Agilent BenchVue, 10
Agilent Technologies, 9–10
Agilent VEE (Visual Engineering
    Environment), 10
AOP. *See* Aspect-Oriented
    Programming
Application-based Software, 12
Application Field Platform, 22
application frameworks, 35
application models, 104
architecture-driven frameworks,
    35
arc method, 195
Aspect-Oriented Programming
    (AOP)
  advantages, 60–62
  crosscutting concerns, 56–58
  join point model, 59
  sample implementation, 59–60
Asynchronous faults, 81

automatic and control systems, 20
automatic magnetic measurement
    systems
  at CERN, 139–140
  flexibility requirements
    hardware overview, 143–144
    past experiences and needs,
      142
    platform at CERN, 142–143
    software requirements, 144
Azimuth DIRECTOR™ II, 10
Azimuth Systems, 10

B

Basic Service Layer
  implementation classes
    Active device, 175–176
    CommunicationBus, 172–175
    CurrentMeter, 176–178
    Cyro-Thermometer, 178–180
    Fast Digital Integrator,
      183–187
    MidiMotorController,
      180–183
    Transducer, 176
  SFMA architecture, 70–71
behavioral patterns, 53
bespoke software, 16
black-box frameworks, 36
brain class, 218–219
brain method, 220

C

CERN. *See* European Organization
for Nuclear Research
class, 50
class Labeled Petri Net
Arc Method, 195
Place Method, 194
Transition Method, 194–195
class structuring, 124
class Synchronizer
Event Node method, 196–197
Node Status method, 197
Task Node method, 196
collaboration disharmonies,
220–221
CommunicationBus, 172–175
computational design pattern, 52
configuration fault methods, 189
core service layer
Fault Detector
configuration fault methods,
189
fatal fault methods, 188–189
local fault methods, 190
warning fault methods,
189–190
interface IFault, 191
SFMA architecture, 71
coupling dispersion, 125
coupling intensity, 125
creational patterns, 53
crosscutting concerns, 56–58
cross references, 206
Cryo-Thermometer, 178–180
CurrentMeter, 176–178

D

Data Class, 218
data-driven frameworks, 35
Data Translation, 12
decoupling, 50
dedicated devices, 21–22
developer/administrator user, 42
development environment, 16
dialogue models, 104

Digital Metrology Solutions, 12
disharmonies, 217
distributed and diagnostics
systems, 18–19
domain frameworks, 35
domain specific design pattern, 54
Domain Specific Languages
(DSL), 38–41
DSL. *See* Domain Specific
Languages
dynamic crosscutting, 59
dynamic dispatch, 50

E

EADS North America Test and
Services, 11
encapsulation, 50
end user, 42
energy control field, 20
European Organization for
Nuclear Research (CERN),
133, 139–140, 142–143
Event Node method, 196–197
event nodes, 92–93
execution pattern, 52
external software quality, 121

F

Fast Digital Integrator (FDI),
183–187
fatal fault methods, 188–189
fault detection software analysis,
227–230
fault detection subsystem, 81
Fault Detector
Flexible Framework For
Magnetic Measurements,
147–152
software framework for
measurement applications
architecture, 81–85
concepts, 80–81
measurement automation,
79–80

validation
discussion, 234–236
fault detection software
analysis, 227–230
measurement procedure, 227
modularity comparison,
230–233
performance verification,
233–234
on rotating coils, 226
fault interception, 83
fault notification publish-subscribe
architecture, 84
fault notification subsystem, 81
FDI. *See* Fast Digital Integrator
Feature Envy problem, 219
Fermi National Accelerator
Laboratory, 139
FFMM. *See* Flexible Framework
for Magnetic Measurements
flexibility, 4–5
flexibility experimental tests
adding/modifying device,
274–276
changing service strategies, 276
description, 272–273
experimental results, 273–274
implementing new measurement
algorithms, 276–277
Flexible Framework for Magnetic
Measurements (FFMM)
Advanced User Interface
Generator
configuration, 160–163
input/output values, 163–164
plotting functions, 164–165
architecture, 146–147
design, 146
FaultDetector, 147–152
Measurement Domain Specific
Language, 156–160
Synchronizer, 152–156
Flexible Measurement Systems
description, 16–17
software frameworks, 17–18

specific hardware, 18
frameworks
application, 35
architecture-driven, 35
classification, 35
data-driven, 35
domain, 35
domain specific languages,
38–41
for measurements, 36–38
requirements for measurement
applications, 41–43
support, 35

G
generally accepted semantics, 118
General Purpose Languages
(GPLs), 40
GIC. *See* Graphic Interaction
Component
God Class, 217–218
GPLs. *See* General Purpose
Languages
grammar language, 204–205
graphical control field, 20
Graphical User Interface engine,
108–109
Graphic Interaction Component
(GIC), 163–164

H
Hall probes, 138
Hardware and Software Platforms
Distributed and Diagnostics
Systems, 18–19
Flexible Measurement Systems
description, 16–17
Software Frameworks, 17–18
Specific Hardware, 18
hardware devices, 80
harmonies, 216
High-Energy Particle (HEP)
accelerators, 134
high-level structuring, 124

I

identifier token, 207
implementation strategy patterns,
    52
information availability, 6
inheritance, 50
instance variable, 50
integrability, 6
intensive coupling, 221–222
interface *IFault*, 191
internal software quality, 121–122
intrinsic operation complexity, 124
ISO 9126 characterization,
    209–214

J

join point model, 59

K

k-times retry, 80

L

Labeled Petri Net (LPN), 91
LabVIEW, 9
LabVIEW SignalExpress, 9
LabWindows/CVI, 9
Large Hadron Collider (LHC),
    134–135
LHC. *See* Large Hadron Collider
LHC tracking test
    background, 265–267
    experimental results, 267–268,
        271–272
    purpose, 264–265
    test procedure, 268–271
Local Fault methods, 190
LPN. *See* Labeled Petri Net

M

Magnetic Field Harmonic
    Analysis, 136–138
Magnetic Field Measurements
    Hall probes, 138
    HEP accelerators, 134

Large Hadron Collider, 134–135
Magnetic Field Harmonic
    Analysis, 136–138
magnetic resonance technique,
    138
rotating coil method, 136
stretched-wire technique, 138
Magnetic Measurement Program
    (MMP), 140–141
magnetic measurement systems
automatic, 139–140
software, 140–141
magnetic permeability
    measurements
    Advanced User Interface
        Generator, 247–249
    Measurement Domain Specific
        Language, 243–246
    On-Field Functional Tests
        background, 253–254
        description, 252–253
        experimental results, 258–259
        experimental set up, 254–255
        FFMM implementation,
            256–258
        test procedure, 255–256
    Petri-net *Synchronizer*, 236–237
magnetic resonance technique, 138
maintainability, 4
MDE. *See* Model-Driven
    Engineering
MDSL. *See* Measurement Domain
    Specific Language
Measurement Domain Specific
    Language (MDSL)
    architecture, 98–99
    builder, 100
    description, 95–97
    discussion, 246–247
    Eclipse platform, 198–202
    editor
        assignment tokens/properties,
            205–206
        code generation with *xPand,*

204
  comments, 208
  cross references, 206
  defining, 208–029
  grammar language, 204–205
  identifier token, 207
  metatype inheritance,
    206–207
  running, 202–204
  type rules, 205
FFMM, 156–160
  goals, 97–98
  magnetic permeability
    measurement, 243–246
  parser, 99–100
  superconducting magnet testing,
    241–243
measurement environment, 80
Measurement Service Layer
  class *Labeled Petri Net*
    arc method, 195
    place method, 194
    transition method, 194–195
  class *Synchronizer*
    event node method, 196–197
    node status method, 197
    task node method, 196
  Petri Net components
    arc, 193
    place, 192
    transition, 192–193
  SFMA architecture, 71
Measurement Studio, 9
message passing, 50
metatype inheritance, 206–207
method, 50
MidiMotorController, 180–183
MMP. *See* Magnetic Measurement
  Program
Model-Driven Engineering
  (MDE), 38
Model-Viewer-Interactor
  paradigm, 102–104
monitoring control field, 20

motion control field, 20
mutex, 86

N
National Instruments, 8
node status method, 197

O
object, 50
Object-Oriented Programming
  advantages, 54–56
  concepts, 47–51
  Patterns
    classification and list, 53–54
    design, 52
    elements, 51–52
On-Field Functional Tests
  LHC tracking test
    background, 265–267
    experimental results, 267–
      268, 271–272
    purpose, 264–265
    test procedure, 268–271
  magnetic permeability
    measurements
    background, 253–254
    description, 252–253
    experimental results, 258–259
    experimental set up, 254–255
    FFMM implementation,
      256–258
    test procedure, 255–256
  rotating coils
    description, 259–260
    experimental results, 262–264
    experimental setup, 261–262
    FFMM implementation, 262
OpenLAB Laboratory Software
  Framework, 10
open recursion, 51
operation structuring, 124

P
parsing, 99–100

PAWS Developers Studio, 11
Petri Net components
  arc, 193
  place, 192
  transition, 192–193
Petri-net *Synchronizer*
  discussion, 239–241
  magnetic permeability
    measurement, 236–237
  measurement procedure,
    237–239
place method, 194
polymorphism, 50
portability, 4
presentation models, 104
probability information, 116
PULSE platform, 12

Q
quality pyramid
  characterization
    Brain Class, 218–219
    Brain Method, 220
    collaboration disharmonies,
      220–221
    Data Class, 218
    disharmonies, 217
    Feature Envy problem, 219
    FFMM source code, 214–216
    God Class, 217–218
    harmonies, 216
    intensive coupling, 221–222
    Shotgun Surgery, 222
  interpretation, 126
  system coupling, 125
  system inheritance, 125
  system size and complexity,
    123–124

R
reusability, 4
reuse software, 6
root nodes, 93
rotating coil method

magnetic field measurements,
  136
On-Field Functional Tests
  description, 259–260
  experimental results, 262–264
  experimental setup, 261–262
  FFMM implementation, 262

S
scalability, 6
sensor networks, 20–21
SFMA. *See* Software Framework
  for Measurement Applications
Shotgun Surgery, 222
SigBase, 11–12
Signal Analysis Software, 12
software components, 80
software environments, 22–23
Software Framework for
  Measurement Applications
  (SFMA)
  architecture, 70–72
  concepts, 69
  design, 72–78
  working principle, 69–70
software magnetic measurement
  systems, 140–141
software market solutions
  criteria for software selection,
    5–6
  market leaders and products,
    6–14
software quality
  assessment
    ISO 9126 characterization,
      209–214
    pyramid characterization,
      214–222
  description, 115–116
  external, 121
  internal, 121–122
  metrics, 116–118
  models, 118–120
  standard ISO 9126, 120–122

in use, 121
software selection criteria
  information availability, 6
  reuse, 6
  scalability, 6
  usability, 5–6
Specific and Custom Software
  automatic and control systems,
    20
  dedicated devices, 21–22
  sensor networks, 20–21
standard ISO 9126, 120–122
static crosscutting, 59
strategy pattern, 52
stretched-wire technique, 138
structural design patterns, 52
structural patterns, 53
structured graphical method, 18
superconducting magnet testing,
  241–243
support frameworks, 35
Synchronizer, 152–156
  concepts, 87–91
  design, 91–93
  evolution example, 93–95
  measurement automation, 85–87
  Petri-Net
    discussion, 239–241
    magnetic permeability
      measurement, 236–237
    measurement procedure,
      237–239
Synchronous faults, 80–81

T
tailor-made software, 16
task node method, 196
task nodes, 93
TestBase, 11

Test Development Software, 12
test engineer, 42
Test Management Software, 12
Test Requirements Document
  (TRD), 11–12
Transducer, 176
transition method, 194–195
transmission control field, 20
TRD. *See* Test Requirements
  Document

U
usability, 5–6
user interfaces
  concepts, 104–105
  Graphical User Interface engine,
    108–109
  interactor, 106–107
  measurement automation,
    101–102
  model-viewer-interactor
    paradigm, 102–104
  view description, 105–106
User Service Layer
  Measurement Domain Specific
    Language
    Eclipse platform, 198–202
    editor, 202–209
  SFMA architecture, 72

W
warning fault methods, 189–190
Web-based distributed and
  diagnostics systems, 18–19
white-box frameworks, 35

Y
Yokogawa, 12

www.ingramcontent.com/pod-product-compliance
Lightning Source LLC
Chambersburg PA
CBHW060328220326
41598CB00023B/2640